Theory of Elasticity

T. G. Sitharam · L. Govindaraju

Theory of Elasticity

 Springer

T. G. Sitharam
Indian Institute of Technology Guwahati
Guwahati, Assam, India

L. Govindaraju
Department of Civil Engineering, UVCE
Bangalore University
Bengaluru, Karnataka, India

ISBN 978-981-33-4652-9   ISBN 978-981-33-4650-5 (eBook)
https://doi.org/10.1007/978-981-33-4650-5

This Springer imprint is published by the registered company Springer Nature Singapore Pte Ltd.
The registered company address is: 152 Beach Road, #21-01/04 Gateway East, Singapore 189721,
Singapore

Foreword I

A course in Theory of Elasticity is a necessity for the postgraduate and senior under-graduate students in Civil and Mechanical Engineering to understand the behaviour of elastic solids under applied loads and also the limitation of the results given by the Elementary Mechanics of Materials. In this text, the authors have developed the subject in detail and in stages in a student-friendly manner. In particular, the large number of problems worked out and a variety of problems given as exercise at the end of each chapter (if worked out) will help the students to gain a good insight into the subject.

I hope the student and teacher community will appreciate the efforts put into this endeavour and make use of this text in their studies and teaching in the area of Solid Mechanics.

K. T. S. Iyengar
Former Professor of Civil Engineering
Indian Institute of Science, Bangalore, India

Foreword II

It is my pleasure to write a foreword to a book authored by my colleagues. The present book contains eight chapters starting with the basic concepts of stress and strain ending with solutions to the problem in geomechanics. The other chapters have clearly brought out the treatment of two-dimensional problems in Cartesian and polar coordinates. The book stands out for its chapter on solutions to problems in geomechanics because such an exclusive chapter dealing with geomechanics is generally not found in the books on elasticity.

I hope that the book fills the gap between the theory and practice. I congratulate the authors on their effort to bring out such a book and I wish them the best in their endeavour.

B. K. Raghu Prasad
Former Professor & Chairman
Department of Civil Engineering
Indian Institute of Science, Bangalore, India

Preface

The aim of writing this textbook is to present the subject contents of applied elasticity course in a clear and logical manner, which constitutes an elective course for senior undergraduate and postgraduate programs in civil, mechanical, aerospace and material engineering departments of many universities. This book highlights the basic principles and the corresponding mathematical expressions involved in theory of elasticity along with applications to many problems in civil engineering including geomechanics. All the logical steps are included for the derivation of the equations and/or solution of all the problems so that the students will be able to follow the textbook very easily. Throughout this book, a simple matrix notation has been adopted in the development of mathematical expressions. A number of solved and unsolved problems have been included in each chapter.

The subject matter has been presented in eight chapters in which Chap. 1 provides a brief introduction to elasticity. Analysis of stress, analysis of strain and stress–strain relations are covered in Chaps. 2–4, respectively. Chapters 5 and 6 cover two-dimensional problems in elasticity in Cartesian coordinate system and polar coordinate system, respectively. Torsion of prismatic bars is covered in Chap. 7. Solutions and applications in geomechanics are presented in Chap. 8.

During the preparation of this book, many books and papers have been referred, and thus, the authors acknowledge all the individuals whose papers and books have been referred. Further, this subject has been offered as a postgraduate course at Indian Institute of Science, Bangalore, emphasizing on the application of elasticity to geotechnical engineering.

We thank Prof. K. T. S. Iyengar, formerly Professor of Civil Engineering, Indian Institute of Science, Bangalore, India, who took pain to go through our manuscript and also for writing a foreword for this book. Also thanks to Prof. B. K. Raghu Prasad, formerly Professor of Civil Engineering, Indian Institute of Science, Bangalore, for all the encouragement and also for writing a foreword for this book. Also thanks to Prof. Gopal Rao, UBDT College of engineering, Davanagere, Karnataka, India, for going through the manuscript and for suggesting many corrections.

We also thank Dr. V. Devaraj, Professor in Civil Engineering, University Visvesvaraya College of Engineering, Bangalore, for his elegant review on this book.

We expect suggestions from the readers both from faculty and from students for the improvement of the book, which will be highly appreciated. Finally, we thank both of our family members for their patience and encouragement.

Guwahati, India   T. G. Sitharam
Bengaluru, India   L. Govindaraju

Review

I thank the authors for having chosen me and have given me an opportunity to review the book titled "Theory of Elasticity". It is needless to re-emphasize the significance and importance of the subject which is mathematical and scary. To use mathematics effectively in applications, you need, *not just knowledge*, but *skill*. You can obtain a certain superficial knowledge of mathematical subjects by listening to lectures, but you can obtain skills only by solving the problems. The authors have bridged the gap between knowledge and skill.

The authors have succeeded in writing a book on "Theory of Elasticity" in a lucid and logical style for the senior undergraduate and postgraduate programs in Civil & Mechanical and Aeronautical Engineering disciplines. In this, the authors have developed the subject matter in a comprehensive manner in stages for the students in a simple and logical way with numerous illustrative examples. The exercise at the end of each chapter helps the student to get an excellent insight into the mathematical subject merits mention.

The authors have presented St Venant's principle which draws the attentions of readers, i.e. "when a body is subjected to various types of loading, the distribution of stress throughout the body is altered only near the region of load application. However the stress distribution is not altered at a distance x=2b irrespective of loading conditions" with illustrative example merits special mention.

The authors have presented torsion of prismatic bar theory and experimental technique of obtaining approximate analytical solutions for bars of narrow sections and members of open thin-walled sections by Prandtl's membrane analogy with illustrative examples and exercise which is complete in all respects.

The two-dimensional problems in polar coordinate systems are illustrated with lot of numerical examples for curved beams. However, the stress concentration factors for sections with circular holes with application to simple tension, compression and bolts would be highly appreciated and had there been an inclusion of numerical examples. The topic of geomechanics is an interesting addition. The application problems on geomechanics with numerical examples are included and then, the authors have accomplished their goal.

I wish the authors on their effort to bring out such an excellent book. I hope the students and teaching fraternity will appreciate the genuine efforts into this endeavour and make use of this in their studies on Theory of Elasticity.

Dr. V. Devaraj
Former Professor of Civil Engineering
University Visvesvaraya College of Engineering
Bangalore University, Bangalore, 560 056 India

Contents

About the Authors

Prof. T. G. Sitharam is currently the Director of Indian Institute of Technology Guwahati (IIT), India and Senior Professor at the Department of Civil Engineering, Indian Institute of Science (IISc), Bengaluru, India. He was a KSIIDC Chair Professor in the area of Energy and Mechanical Sciences and the founder Chairman of the Center for Infrastructure, Sustainable Transport and Urban Planning (CiSTUP) at IISc. He was the Chairman of the AICTE South Western Zonal Committee, Bengaluru and now he is for Eastern zonal committee. He is the President of the Indian Society for Earthquake Technology (ISET) and the founder President of the International Association for Coastal Reservoir Research (IACRR). He was a Visiting Professor at Yamaguchi University, Japan; University of Waterloo, Canada; University of Dolhousie, Halifax, Canada; and ISM Dhanbad, India. Was a Research Scientist at University of Texas at Austin, Texas, USA.

Dr. L. Govindraju is currently Professor in the Department of Civil Engineering, University Visvesvaraya College of Engineering (UVCE), Bangalore University, Bangalore. He obtained his Bachelor's in Civil Engineering from Mysore University, India (1986), Master's in Geotechnical Engineering from National Institute of Technology Karnataka (NITK), India (1994) and Ph.D. in Civil Engineering from Indian Institute of Science (IISc), India (2005). He was awarded fellowship from National Programme on Earthquake Engineering Education (NPEEE) instituted by the Ministry of Human Resources and Development, Government of India to pursue post doctoral research in the area of earthquake engineering at the Department of Engineering Science, University of Oxford (UK). He is the member of several professional bodies such as IGS, ISTE and ISET. He has published over 150 technical papers and two books.

Chapter 1
Elasticity

1.1 Introduction

Elasticity is a fascinating subject that deals with determination of stresses, strains and displacement distribution in an elastic body under the influence of external forces. If the external forces producing deformation do not exceed a certain limit, the deformation disappears with the removal of the forces. Thus, the elastic behaviour implies the absence of any permanent deformation. Elasticity has been developed following the great achievement of Newton in stating the laws of motion, although it has earlier roots. The need to understand and control the fracture of solids seems to have been a first motivation. Leonardo da Vinci sketched in his notebooks a possible test of the tensile strength of a wire. Galileo had investigated the breaking loads of rods under tension and concluded that the load was independent of length and proportional to the cross-sectional area, this being the first step towards a concept of stress.

Every engineering material possesses a certain extent of elasticity. The common materials of construction would remain elastic only for very small strains before exhibiting either plastic straining or brittle failure. However, natural polymeric materials show elasticity over a wider range (usually with time or rate effects, thus they would more accurately be characterized as viscoelastic), and the widespread use of natural rubber and similar materials motivated the development of finite elasticity. While many roots of the subject were laid in the classical theory, especially in the work of Green, Gabrio Piola and Kirchhoff in the mid-1800s, the development of a viable theory with forms of stress–strain relations for specific rubbery elastic materials, as well as an understanding of the physical effects of the nonlinearity in simple problems such as torsion and bending, was mainly the achievement of the British-born engineer and applied mathematician Ronald S. Rivlin in the 1940s and 1950s.

© The Author(s), under exclusive license to Springer Nature Singapore Pte Ltd. 2021
T. G. Sitharam and L. Govindaraju, *Theory of Elasticity*,
https://doi.org/10.1007/978-981-33-4650-5_1

1.2   The General Theory of Elasticity

Linear elasticity as a general three-dimensional theory has been developed in the early 1820s based on Cauchy's work. Simultaneously, Navier had developed an elasticity theory based on a simple particle model, in which particles interacted with their neighbours by a central force of attraction between neighbouring particles. Later, it was gradually realized, following work by Navier, Cauchy and Poisson in the 1820s and 30s, the particle model is too simple. Most of the subsequent development of this subject was in terms of the continuum theory. George Green highlighted the maximum possible number of independent elastic moduli in the most general anisotropic solid in 1837. Green pointed out that the existence of elastic strain energy required that of the 36 elastic constants relating the 6 stress components to the 6 strains, at most 21 could be independent. In 1855, Lord Kelvin showed that a strain energy function must exist for reversible isothermal or adiabatic response and showed that temperature changes are associated with adiabatic elastic deformation. The middle and late 1800s were a period in which many basic elastic solutions were derived and applied to technology and to the explanation of natural phenomena. Adhémar-Jean-Claude Barré de Saint–Venant derived in the 1850s solutions for the torsion of noncircular cylinders, which explained the necessity of warping displacement of the cross-section in the direction parallel to the axis of twisting, and for the flexure of beams due to transverse loading; the latter allowed understanding of approximations inherent in the simple beam theory of Jakob Bernoulli, Euler and Coulomb. Heinrich Rudolf Hertz developed solutions for the deformation of elastic solids as they are brought into contact and applied these to model details of impact collisions. Solutions for stress and displacement due to concentrated forces acting at an interior point of a full space were derived by Kelvin and those on the surface of a half space by Boussinesq and Cerruti. In 1863, Kelvin had derived the basic form of the solution of the static elasticity equations for a spherical solid, and these were applied in following years to such problems as calculating the deformation of the Earth due to rotation and tidal forcing and measuring the effects of elastic deformability on the motions of the Earth's rotation axis. The classical development of elasticity never fully confronted the problem of finite elastic straining, in which material fibres change their lengths by other than very small amounts.

1.3   Assumptions of Linear Elasticity

In order to evaluate the stresses, strains and displacements in an elasticity problem, one needs to derive a series of basic equations and boundary conditions. During the process of deriving such equations, one can consider all the influential factors, the results obtained will be so complicated, and hence practically no solutions can be found. Therefore, some basic assumptions have to be made about the properties of the body considered to arrive at possible solutions. Under such assumptions, we can neglect some of the influential factors of minor importance. The following are the assumptions in classical elasticity.

The Body is Continuous
Here the whole volume of the body is considered to be filled with continuous matter, without any void. Only under this assumption, can the physical quantities in the body, such as stresses, strains and displacements, be continuously distributed and thereby expressed by continuous functions of coordinates in space. However, these assumptions will not lead to significant errors so long as the dimensions of the body are very large in comparison with those of the particles and with the distances between neighbouring particles.

The Body is Perfectly Elastic
The body is considered to be wholly obeys Hooke's law of elasticity, which shows the linear relations between the stress components and strain components. Under this assumption, the elastic constants will be independent of the magnitudes of the stress and strain components.

The Body is Homogenous
In this case, the elastic properties are the same throughout the body. Thus, the elastic constants will be independent of the location in the body. Under this assumption, one can analyse an elementary volume isolated from the body and then apply the results of analysis to the entire body.

The Body is Isotropic
Here the elastic properties in a body are the same in all directions. Hence, the elastic constants will be independent of the orientation of coordinate axes.

The Displacements and Strains are Small
The displacement components of all points of the body during deformation are very small in comparison with its original dimensions, and the strain components and the rotations of all line elements are much smaller than unity. Hence, when formulating the equilibrium equations relevant to the deformed state, the lengths and angles of the body before deformation are used. In addition, when geometrical equations involving strains and displacements are formulated, the squares and products of the small quantities are neglected. Therefore, these two measures are necessary to linearize the algebraic and differential equations in elasticity for their easier solution.

1.4   Applications of Linear Elasticity

The very purpose of application of elasticity is to analyse the stresses and displacements of elements within the elastic range and thereby to check the sufficiency of their strength, stiffness and stability.

Civil engineering applications involve important contributions to stress and deflection analysis of structures, such as beams, plates, shells and rods. Further applications include in geomechanics involving the stresses in materials such as soil, rock, concrete and asphalt.

Mechanical engineering includes application of elasticity in many problems such as in the analysis and design of machine elements. These applications include analysis of general stresses distribution in solids, contact stresses, thermal stresses, fatigue and fracture mechanics.

Aerospace and aeronautical engineering uses elasticity in the analysis of stress, fracture and fatigue analysis in aerostructures.

Materials engineering applies elasticity in the determination of stress fields in crystalline solids, s around dislocations and in materials with microstructure.

Although, elasticity, mechanics of materials and structural mechanics are the three branches of solid mechanics, they differ from one to other both in the objects and methods of analysis.

Mechanics of materials deals essentially with the stresses and displacements of a structural or machine element in the shape of a bar, straight or curved, which is subjected to tension, compression, shear, bending or torsion. Structural mechanics, on the basis of mechanics of materials, deals with the stresses and displacements of a structure in the form of a bar system, such as a truss or a rigid frame. As to the structural elements that are not in form of a bar, such as blocks, plates, shells, dams and foundations, they are analysed only in the theory of elasticity. Moreover, in order to analyse a bar element thoroughly and very precisely, it is necessary to apply the theory of elasticity.

Although bar-shaped elements are studied both in mechanics of materials and in theory of elasticity, the methods of analysis used here are not entirely the same. When the element is studied in mechanics of materials, some assumptions are usually made on the strain condition or the stress distribution. These assumptions simplify the mathematical derivation to a certain extent, but many a times inevitably reduce the degree of accuracy of the results obtained. However, in elasticity, the study of bar-shaped element usually does not need those assumptions. Thus, the results obtained by the application of elasticity theory are more accurate and may be used to check the appropriate results obtained in mechanics of materials.

While analysing the problems of bending of straight beam under transverse loads by the mechanics of materials, it is usual to assume that a plane section before bending of the beam remains plane even after the bending. This assumption leads to the linear distribution of bending stresses. In the theory of elasticity, however one can solve the problem without this assumption and prove that the stress distribution will be far from linear variation as shown in the next sections.

Further, while analysing for the distribution of stresses in a tension member with a hole, it is assumed in mechanics of materials that the tensile stresses are uniformly distributed across the net section of the member, whereas the exact analysis in the theory of elasticity shows that the stresses are by no means uniform, but are concentrated near the hole; the maximum stress at the edge of the hole is far greater than the average stress across the net section.

The theory of elasticity contains equilibrium equations relating to the stresses; kinematic equations relating to the strains and displacements; constitutive equations relating to the stresses and strains; boundary conditions relating to the physical domain; and uniqueness constraints relating to the applicability of the solution.

Chapter 2
Analysis of Stress

2.1 Introduction

A body under the action of external forces undergoes distortion, and the effect due to this system of forces is transmitted throughout the body developing internal forces in it. To examine these internal forces at a point O in Fig. 2.1a, inside the body, consider a plane MN passing through the point O. If the plane is divided into a number of small areas, as in Fig. 2.1b, and the forces acting on each of these measured, it will be observed that these forces vary from one small area to the next. On the small area ΔA at point O, there will be acting a force of ΔF as shown in Fig. 2.1b. From this, it can be understood the concept of stress as the internal force per unit area. Assuming the material is continuous, the term "stress" at any point across a small area ΔA can be defined by the limiting Equation as below.

$$\text{Stress} = \lim_{\Delta A \to 0} \frac{\Delta F}{\Delta A} \tag{2.1}$$

where ΔF is the internal force on the area ΔA surrounding the given point. Stress is sometimes referred to as force intensity.

2.2 Notation of Stress

Here, a single suffix σ notation, like $\sigma_x, \sigma_y, \sigma_z$, is used for the direct stresses and double suffix τ notation is used for shear stresses like τ_{xy}, τ_{xz}, etc. τ_{xy} means a stress, produced by an internal force in the direction of y, acting on a surface, having a normal in the direction of x.

© The Author(s), under exclusive license to Springer Nature Singapore Pte Ltd. 2021
T. G. Sitharam and L. Govindaraju, *Theory of Elasticity*,
https://doi.org/10.1007/978-981-33-4650-5_2

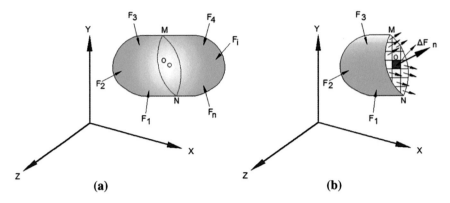

Fig. 2.1 Forces acting on a body

2.3 Concept of Direct Stress and Shear Stress

Figure 2.2 shows the rectangular components of the force vector ΔF referred to corresponding axes. Taking the ratios $\frac{\Delta F_x}{\Delta A_x}$, $\frac{\Delta F_y}{\Delta A_x}$, $\frac{\Delta F_z}{\Delta A_x}$, we have three quantities that establish the average intensity of the force on the area ΔA_x. In the limit as $\Delta A \rightarrow 0$, the above ratios define the force intensity acting on the x-face at point O. These values of the three intensities are defined as the "Stress components" associated with the x-face at point O. The stress components parallel to the surface are called "shear stress components" denoted by τ. The shear stress component acting on the x-face in the y-direction is identified as τ_{xy}. The stress component perpendicular to the face is called "normal stress" or "direct stress" component and is denoted by σ. This is identified as σ_x along x-direction. From the above discussions, the stress components on the x-face at point O are defined as follows in terms of force intensity ratios

Fig. 2.2 Force components of ΔF acting on small area centred on point O

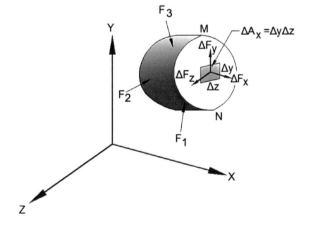

Fig. 2.3 Stress components at point O

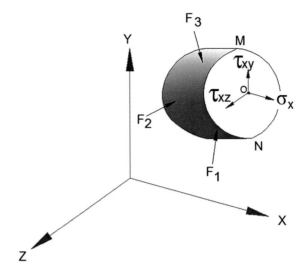

$$\sigma_x = \lim_{\Delta A_x \to 0} \frac{\Delta F_x}{\Delta A_x}$$

$$\tau_{xy} = \lim_{\Delta A_x \to 0} \frac{\Delta F_y}{\Delta A_x}$$

$$\tau_{xz} = \lim_{\Delta A_x \to 0} \frac{\Delta F_z}{\Delta A_x} \tag{2.2}$$

The above stress components are illustrated in Fig. 2.3.

2.4   Stress Tensor

Let O be the point in a body shown in Fig. 2.1a. Passing through that point, infinitely many planes may be drawn. As the resultant forces acting on these planes are the same, the stresses on these planes are different because the areas and the inclinations of these planes are different. Therefore, for a complete description of stress, we have to specify not only its magnitude, direction and sense but also the surface on which it acts. For this reason, the stress is called a "Tensor".

Figure 2.4 depicts three-orthogonal co-ordinate planes representing a parallelopiped on which are nine components of stress. Of these, three are direct stresses and six shear stresses. In tensor notation, these can be expressed by the tensor τ_{ij}, where $i = x, y, z$ and $j = x, y, z$. In matrix notation, it is often written as

$$\tau_{ij} = \begin{bmatrix} \tau_{xx} & \tau_{xy} & \tau_{xz} \\ \tau_{yx} & \tau_{yy} & \tau_{yz} \\ \tau_{zx} & \tau_{zy} & \tau_{zz} \end{bmatrix} \tag{2.3}$$

Fig. 2.4 Stress components acting on parallelopiped

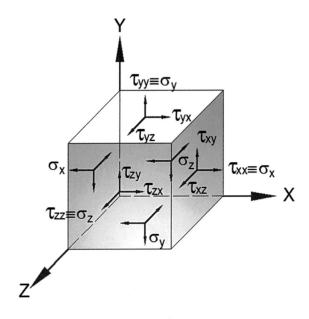

otherwise, it is written as

$$S = \begin{bmatrix} \sigma_x & \tau_{xy} & \tau_{xz} \\ \tau_{yx} & \sigma_y & \tau_{yz} \\ \tau_{zx} & \tau_{zy} & \sigma_z \end{bmatrix}$$

(2.4)

if we use ordinary expression in matrix form.

2.5 Spherical and Deviatorial Stress Tensors

A general stress tensor can be conveniently divided into two parts as shown above. Let us now define a new stress term (σ_m) as the mean stress, so that

$$\sigma_m = \frac{\sigma_x + \sigma_y + \sigma_z}{3}$$

(2.5)

Imagine a hydrostatic type of stress having all the normal stresses equal to σ_m, and all the shear stresses are zero. We can divide the stress tensor into two parts, one having only the "hydrostatic stress" and the other, "deviatorial stress". The hydrostatic type of stress is given by

$$\begin{bmatrix} \sigma_m & 0 & 0 \\ 0 & \sigma_m & 0 \\ 0 & 0 & \sigma_m \end{bmatrix} \tag{2.6}$$

The deviatorial type of stress is given by

$$\begin{bmatrix} \sigma_x - \sigma_m & \tau_{xy} & \tau_{xz} \\ \tau_{xy} & \sigma_y - \sigma_m & \tau_{yz} \\ \tau_{xz} & \tau_{yz} & \sigma_z - \sigma_m \end{bmatrix} \tag{2.7}$$

Here, the hydrostatic type of stress is known as "spherical stress tensor", and the other is known as the "deviatorial stress tensor".

It will be seen later that the deviatorial part produces changes in shape of the body and finally causes failure. The spherical part is rather harmless, produces only uniform volume changes without any change of shape and does not necessarily cause failure.

2.6 Indicial Notation

An alternate notation called index or indicial notation for stress is more convenient for general discussions in elasticity. In indicial notation, the co-ordinate axes x, y and z are replaced by numbered axes x_1, x_2 and x_3, respectively. The components of the force ΔF of Fig. 2.1a are written as ΔF_1, ΔF_2 and ΔF_3, where the numerical subscript indicates the component with respect to the numbered co-ordinate axes.

The definitions of the components of stress acting on the x_1 face can be written in indicial form as follows:

$$\sigma_{11} = \lim_{\Delta A_1 \to 0} \frac{\Delta F_1}{\Delta A_1}$$
$$\sigma_{12} = \lim_{\Delta A_1 \to 0} \frac{\Delta F_2}{\Delta A_1}$$
$$\sigma_{13} = \lim_{\Delta A_1 \to 0} \frac{\Delta F_3}{\Delta A_1} \tag{2.8}$$

Here, the symbol σ is used for both normal and shear stresses.

In general, all components of stress can now be defined by a single equation:

$$\sigma_{ij} = \lim_{\Delta A_i \to 0} \frac{\Delta F_j}{\Delta A_i} \tag{2.9}$$

Here, i and j take on the values 1, 2 and 3.

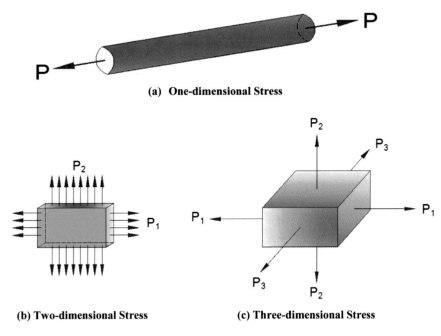

(a) One-dimensional Stress

(b) Two-dimensional Stress   **(c) Three-dimensional Stress**

Fig. 2.5 Types of stress

2.7 Types of Stress

Stresses may be classified in two ways, i.e. according to the type of body on which they act, or the nature of the stress itself. Thus, stresses could be one-dimensional, two-dimensional or three-dimensional as shown in Fig. 2.5.

2.8 Two-Dimensional Stress at a Point

A two-dimensional state of stress exists when the stresses and body forces are independent of one of the co-ordinates. Such a state is described by stresses σ_x, σ_y and τ_{xy} and the X and Y body forces (here, z is taken as the independent co-ordinate axis).

We shall now determine the equations for transformation of the stress components σ_x, σ_y and τ_{xy} at any point of a body represented by infinitesimal element as shown in Fig. 2.6.

Consider an infinitesimal wedge cut from the loaded body shown in Fig. 2.6. It is required to determine the stresses $\sigma_{x'}$ and $\tau_{x'y'}$, that refer to axes x', y' making an angle θ with axes x, y as shown in Fig. 2.7. Let side MN be normal to the x' axis.

Fig. 2.6 Thin body subjected to stresses in xy plane

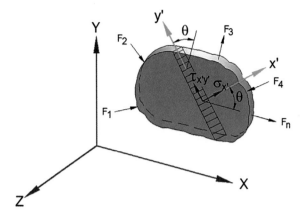

Fig. 2.7 Stress components acting on faces of a small wedge cut from body of Fig. 2.6

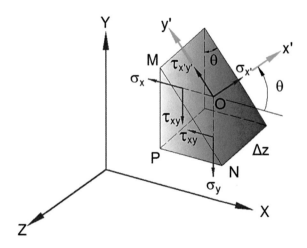

Considering $\sigma_{x'}$ and $\tau_{x'y'}$ as positive and area of side MN as unity, the sides MP and PN have areas $\cos\theta$ and $\sin\theta$, respectively.

Equilibrium of the forces in the x- and y-directions requires that

$$T_x = \sigma_x \cos\theta + \tau_{xy} \sin\theta$$
$$T_y = \tau_{xx} \cos\theta + \sigma_y \sin\theta \qquad (2.10)$$

where T_x and T_y are the components of stress resultant acting on MN in the x- and y-directions, respectively. The normal and shear stresses on the x' plane (MN plane) are obtained by projecting T_x and T_y in the x'- and y'-directions.

$$\sigma_{x'} = T_x \cos\theta + T_y \sin\theta$$
$$\tau_{x'y'} = T_y \cos\theta - T_x \sin\theta \qquad (2.11)$$

Upon substitution of stress resultants from Eq. (2.9), Eq. (2.10) become

$$\sigma_{x'} = \sigma_x \cos^2 \theta + \sigma_y \sin^2 \theta + 2\tau_{xy} \sin \theta \cos \theta$$
$$\tau_{x'y'} = \tau_{xy}\left(\cos^2 \theta - \sin^2 \theta\right) + \left(\sigma_y - \sigma_x\right) \sin \theta \cos \theta \qquad (2.12)$$

The stress $\sigma_{y'}$ is obtained by substituting $\left(\theta + \frac{\pi}{2}\right)$ for θ in the expression for $\sigma_{x'}$. By means of trigonometric identities

$$\cos^2 \theta = \tfrac{1}{2}(1 + \cos 2\theta), \ \sin \theta \cos \theta = \tfrac{1}{2} \sin 2\theta,$$
$$\sin^2 \theta = \tfrac{1}{2}(1 - \cos 2\theta) \qquad (2.13)$$

The transformation equations for stresses are now written in the following form:

$$\sigma_{x'} = \frac{1}{2}\left(\sigma_x + \sigma_y\right) + \frac{1}{2}\left(\sigma_x - \sigma_y\right) \cos 2\theta + \tau_{xy} \sin 2\theta \qquad (2.13a)$$

$$\sigma_{y'} = \frac{1}{2}\left(\sigma_x + \sigma_y\right) - \frac{1}{2}\left(\sigma_x - \sigma_y\right) \cos 2\theta - \tau_{xy} \sin 2\theta \qquad (2.13b)$$

$$\tau_{x'y'} = -\frac{1}{2}\left(\sigma_x - \sigma_y\right) \sin 2\theta + \tau_{xy} \cos 2\theta \qquad (2.13c)$$

2.9 Principal Stresses in Two Dimensions

To ascertain the orientation of $x'y'$ corresponding to maximum or minimum $\sigma_{x'}$, the necessary condition, $\frac{d\sigma_{x'}}{d\theta} = 0$, is applied to Eq. (2.12a), yielding

$$-\left(\sigma_x - \sigma_y\right) \sin 2\theta + 2\tau_{xy} \cos 2\theta = 0 \qquad (2.14)$$

Therefore,

$$\tan \theta = \frac{2\tau_{xy}}{\sigma_x - \sigma_y} \qquad (2.15)$$

As $2\theta = \tan(\pi + 2\theta)$, two directions, mutually perpendicular, are found to satisfy Eq. (2.14). These are the principal directions, along which the principal or maximum and minimum normal stresses act.

When Eq. (2.13c) is compared with Eq. (2.14), it becomes clear that $\tau_{x'y'} = 0$ on a principal plane. A principal plane is thus a plane of zero shear. The principal stresses are determined by substituting Eq. (2.15) into Eq. (2.13a)

$$\sigma_{1,2} = \frac{\sigma_x + \sigma_y}{2} \pm \sqrt{\left(\frac{\sigma_x - \sigma_y}{2}\right)^2 + \tau_{xy}^2} \tag{2.16}$$

Algebraically, larger stress given above is the maximum principal stress, denoted by σ_1. The minimum principal stress is represented by σ_2.

Similarly, by using the above approach and employing Eq. (2.13c), an expression for the maximum shear stress can be expressed as:

$$\text{Maximum shear stress, } \tau_{max} = \left(\frac{\sigma_x - \sigma_y}{2}\right) \tag{2.17}$$

2.10 Cauchy's Stress Principle

According to the general theory of stress by Cauchy (1823), the stress principle can be stated as follows:

Consider any closed surface ΔS within a continuum of region R that separates the region R into sub-regions R_1 and R_2. The interaction between these sub-regions can be represented by a field of stress vectors $T(\hat{n})$ defined on ΔS. By combining this principle with Euler's equations that express balance of linear momentum and moment of momentum in any kind of body, Cauchy derived the following relationship.

$$T(\hat{n}) = -T(-\hat{n})$$
$$T(\hat{n}) = \sigma^T(\hat{n}) \tag{2.18}$$

where $(\hat{n})$ is the unit normal to ΔS and σ is the stress matrix. Further, in the regions where the field variables have sufficiently smooth variations to allow spatial derivatives up to any order, we have

$$\rho A = div\,\sigma + f \tag{2.19}$$

where

$\rho =$ material mass density
$A =$ acceleration field
$f =$ body force per unit volume

This result expresses a necessary and sufficient condition for the balance of linear momentum. When Eq. (2.18) is satisfied,

$$\sigma = \sigma^T \tag{2.20}$$

which is equivalent to the balance of moment of momentum with respect to an arbitrary point. In deriving (2.19), it is implied that there are no body couples. If body couples and/or couple stresses are present, Eq. (2.20) is modified, but Eq. (2.19) remains unchanged.

Cauchy stress principle has four essential ingradients

1. The physical dimensions of stress are (force)/(area).
2. Stress is defined on an imaginary surface that separates the region under consideration into two parts.
3. Stress is a vector or vector field equipollent to the action of one part of the material on the other.
4. The direction of the stress vector is not restricted.

2.11 Direction Cosines

Consider a plane ABC having an outward normal "n" as shown in Fig. 2.8. The direction of this normal can be defined in terms of direction cosines. Let the angle of inclinations of the normal with x, y and z axes be α, β and γ, respectively. Let $P(x, y, z)$ be a point on the normal at a radial distance r from the origin O.

From Fig. 2.8,

$$\cos\alpha = \frac{x}{r}, \quad \cos\beta = \frac{y}{r} \text{ and } \cos\gamma = \frac{z}{r}$$

or

$$x = r\cos\alpha, \quad y = r\cos\beta \text{ and } z = r\cos\gamma$$

Fig. 2.8 Tetrahedron with arbitrary plane

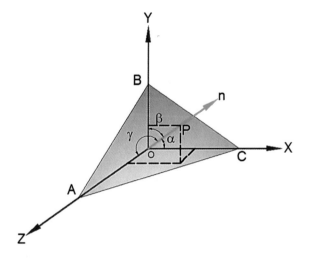

Let

$$\cos\alpha = l, \quad \cos\beta = m \text{ and } \cos\gamma = n$$

Therefore,

$$\frac{x}{r} = l, \quad \frac{y}{r} = m \text{ and } \frac{z}{r} = n$$

Here, l, m and n are known as direction cosines of the line OP. Also, it can be written as

$$x^2 + y^2 + z^2 = r^2 \text{ (since } r \text{ is the polar co-ordinate of } P)$$

or

$$\frac{x^2}{r^2} + \frac{y^2}{r^2} + \frac{z^2}{r^2} = 1$$

$l^2 + m^2 + n^2 = 1$ (This is well known in co-ordinate geometry)

2.12 Stress Components on an Arbitrary Plane

Consider a small tetrahedron isolated from a continuous medium (Fig. 2.9) subjected to a general state of stress. The body forces are taken to be negligible. Let the arbitrary plane ABC be identified by its outward normal n whose direction cosines are l, m and n.

Fig. 2.9 Stresses acting on face of the tetrahedron

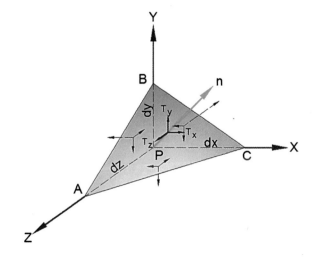

In Fig. 2.9, T_x, T_y, T_z are the Cartesian components of stress resultant T, acting on oblique plane ABC. It is required to relate the stresses on the perpendicular planes intersecting at the origin to the normal and shear stresses acting on ABC.

The orientation of the plane ABC may be defined in terms of the angle between a unit normal n to the plane and the x-, y- and z-directions. The direction cosines associated with these angles are

$$\cos(n, x) = l$$
$$\cos(n, y) = m \text{ and} \quad\quad\quad\quad\quad\quad (2.21)$$
$$\cos(n, z) = n$$

But the three direction cosines are related by

$$l^2 + m^2 + n^2 = 1 \quad\quad\quad\quad\quad\quad (2.22)$$

The area of the perpendicular plane PAB, PAC, PBC may now be expressed in terms of A, the area of ABC, and the direction cosines.

Therefore,

$$\text{Area of PAB} = A_{PAB} = A_x = A.i$$
$$= A(li + mj + nk).i$$
$$\text{Hence, } A_{PAB} = Al$$

The other two areas are similarly obtained. In doing so, we have altogether

$$A_{PAB} = Al, A_{PAC} = Am \text{ and } A_{PBC} = An \quad\quad\quad\quad (2.23)$$

Here, i, j and k are unit vectors in x-, y- and z-directions, respectively.

Now, for equilibrium of the tetrahedron, the sum of forces in x-, y- and z-directions must be zero.

Therefore,

$$T_x A = \sigma_x Al + \tau_{xy} Am + \tau_{xz} An \qu\quad\quad\quad\quad (2.24)$$

Dividing throughout by A, we get

$$T_x = \sigma_x l + \tau_{xy} m + \tau_{xz} n \quad\quad\quad\quad\quad (2.24a)$$

Similarly, for equilibrium in y- and z-directions,

$$T_y = \tau_{xy} l + \sigma_y m + \tau_{yz} n \quad\quad\quad\quad\quad (2.24b)$$

and

$$T_z = \tau_{xz}\, l + \tau_{yz}\, m + \sigma_z\, n \qquad (2.24c)$$

The stress resultant on A is thus determined on the basis of known stresses $\sigma_x, \sigma_y, \sigma_z, \tau_{xy}, \tau_{yz}, \tau_{zx}$ and a knowledge of the orientation of A.

Equations (2.24a), (2.24b) and (2.24c) are known as Cauchy's stress formula. These equations show that the nine rectangular stress components at point P will enable one to determine the stress components on any arbitrary plane passing through point P.

2.13 Stress Components on Oblique Plane (Stress Transformation)

When the state or stress at a point is specified in terms of the six components with reference to a given co-ordinate system, then for the same point, the stress components with reference to another co-ordinate system obtained by rotating the original axes can be determined using the direction cosines.

Consider a cartesian co-ordinate system x, y and z as shown in Fig. 2.10. Let this given co-ordinate system is rotated to a new co-ordinate system X', Y', Z' where in X' lie on an oblique plane. The X', Y', Z' and X, Y, Z systems are related by the direction cosines.

$$
\begin{aligned}
l_1 &= \cos(X', X) \\
m_1 &= \cos(X', Y) \qquad (2.25) \\
n_1 &= \cos(X', Z)
\end{aligned}
$$

Fig. 2.10 Transformation of co-ordinates

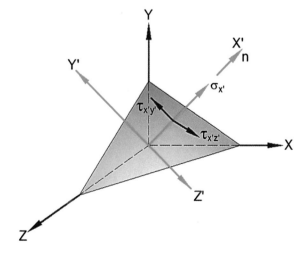

Table 2.1 Direction cosines for transformed co-ordinates

	X	Y	Z
X'	l_1	m_1	n_1
Y'	l_2	m_2	n_2
Z'	l_3	m_3	n_3

(The notation corresponding to a complete set of direction cosines is shown in Table 2.1).

The normal stress $\sigma_{x'}$ is found by projecting T_x, T_y and T_z in the X' direction and adding:

$$\sigma_{x'} = T_x l_1 + T_y m_1 + T_z n_1 \tag{2.26}$$

Equations (2.24a), (2.24b), (2.24c) and (2.26) are combined to yield

$$\sigma_{x'} = \sigma_x l_1^2 + \sigma_y m_1^2 + \sigma_z n_1^2 + 2\left(T_{xy} l_1 m_1 + \tau_{yz} m_1 n_1 + \tau_{xz} l_1 n_1\right) \tag{2.27}$$

Similarly by projecting T_x, T_y, T_z in the y' and z' directions, we obtain, respectively

$$\tau_{x'y'} = \sigma_x l_1 l_2 + \sigma_y m_1 m_2 + \sigma_z n_1 n_2 + \tau_{xy}(l_1 m_2 + m_1 l_2)$$
$$+ \tau_{yz}(m_1 n_2 + n_1 m_2) + \tau_{xz}(n_1 l_2 + l_1 n_2) \tag{2.27a}$$

$$\tau_{x'z'} = \sigma_x l_1 l_3 + \sigma_y m_1 m_3 + \sigma_z n_1 n_3 + \tau_{xy}(l_1 m_3 + m_1 l_3)$$
$$+ \tau_{yz}(m_1 n_3 + n_1 m_3) + \tau_{xz}(n_1 l_3 + l_1 n_3) \tag{2.27b}$$

Recalling that the stresses on three mutually perpendicular planes are required to specify the stress at a point (one of these planes being the oblique plane in question), the remaining components are found by considering those planes perpendicular to the oblique plane. For one such plane n would now coincide with y' direction, and expressions for the stresses $\sigma_{y'}$, $\tau_{y'}$, $\tau_{y'z'}$ would be derived. In a similar manner, the stresses $\sigma_{z'}$, $\tau_{z'x'}$, $\tau_{z'y'}$ are determined when n coincides with the z' direction. Owing to the symmetry of stress tensor, only six of the nine stress components thus developed are unique. The remaining stress components are as follows:

$$\sigma_{y'} = \sigma_x l_2^2 + \sigma_y m_2^2 + \sigma_z n_2^2 + 2\left(\tau_{xy}\, l_2\, m_2 + \tau_{yz}\, m_2\, n_2 + \tau_{xz}\, l_2\, n_2\right) \tag{2.27c}$$

$$\sigma_{z'} = \sigma_x l_3^2 + \sigma_y m_3^2 + \sigma_z n_3^2 + 2\left(\tau_{xy}\, l_3\, m_3 + \tau_{yz}\, m_3\, n_3 + \tau_{xz}\, l_3\, n_3\right) \tag{2.27d}$$

$$\tau_{y'z'} = \sigma_x l_2 l_3 + \sigma_y m_2 m_3 + \sigma_z n_2 n_3 + \tau_{xy}(m_2 l_3 + l_2 m_3)$$
$$+ \tau_{yz}(n_2 m_3 + m_2 n_3) + \tau_{xz}(l_2 n_3 + n_2 l_3) \tag{2.27e}$$

Equations (2.27)–(2.27e) represent expressions transforming the quantities $\sigma_x, \sigma_y, \tau_{xy}, \tau_{yz}, \tau_{xz}$ to completely define the state of stress.

It is to be noted that because, X', Y' and Z' are orthogonal, the nine direction cosines must satisfy trigonometric relations of the following form.

$$l_i^2 + m_i^2 + n_i^2 = 1 \quad (i = 1, 2, 3)$$

and

$$
\begin{aligned}
l_1 l_2 + m_1 m_2 + n_1 n_2 &= 0 \\
l_2 l_3 + m_2 m_3 + n_2 n_3 &= 0 \\
l_1 l_3 + m_1 m_3 + n_1 n_3 &= 0
\end{aligned}
\tag{2.27f}
$$

If we denote the expressions for direction cosines given in Table 2.1 by the matrix $[a]$, then the nine stress components in the new co-ordinate system X', Y', Z', can be written as

$$\left[\sigma'\right] = [a][\sigma][a]^T \tag{2.27f}$$

In other words, in an expanded form:

$$
\begin{bmatrix}
\sigma_{x'} & \tau_{x'y'} & \tau_{x'z'} \\
\tau_{y'x'} & \sigma_{y'} & \tau_{y'z'} \\
\tau_{z'x'} & \tau_{z'y'} & \sigma_{z'}
\end{bmatrix}
=
\begin{bmatrix}
l_1 & m_1 & n_1 \\
l_2 & m_2 & n_2 \\
l_3 & m_3 & n_3
\end{bmatrix}
\begin{bmatrix}
\sigma_x & \tau_{xy} & \tau_{xz} \\
\tau_{yx} & \sigma_y & \tau_{yz} \\
\tau_{zx} & \tau_{zy} & \sigma_z
\end{bmatrix}
\begin{bmatrix}
l_1 & l_2 & l_3 \\
m_1 & m_2 & m_3 \\
n_1 & n_2 & n_3
\end{bmatrix}
\tag{2.27h}
$$

2.14 Principal Stress in Three Dimensions

For the three-dimensional case, it is required that three planes of zero shear stress exist, that these planes are mutually perpendicular, and that on these planes, the normal stresses have maximum or minimum values. As discussed earlier, these normal stresses are referred to as principal stresses, usually denoted by σ_1, σ_2 and σ_3. The largest stress is represented by σ_1 and the smallest by σ_3.

Again considering an oblique plane X', the normal stress acting on this plane is given by Eq. (2.27).

$$\sigma_{x'} = \sigma_x l^2 + \sigma_y m^2 + \sigma_z n^2 + 2(\tau_{xy} \, lm + \tau_{yz} \, mn + \tau_{xz} \, ln) \tag{2.28}$$

The problem here is to determine the extreme or stationary values of $\sigma_{x'}$. To accomplish this, we examine the variation of $\sigma_{x'}$ relative to the direction cosines. As l, m and n are not independent, but connected by $l^2 + m^2 + n^2 = 1$, only l and m may be regarded as independent variables.

Thus,

$$\frac{\partial \sigma_{x'}}{\partial l} = 0, \quad \frac{\partial \sigma_{x'}}{\partial m} = 0 \tag{2.28a}$$

Differentiating Eq. (2.28), in terms of the quantities in Eqs. (2.24a), (2.24b) and (2.24c), we obtain

$$\begin{aligned} T_x + T_z \frac{\partial n}{\partial l} &= 0, \\ T_y + T_z \frac{\partial n}{\partial m} &= 0, \end{aligned} \tag{2.28b}$$

From $n^2 = 1 - l^2 - m^2$, we have

$$\frac{\partial n}{\partial l} = -\frac{l}{n} \text{ and } \frac{\partial n}{\partial m} = -\frac{m}{n}$$

Introducing the above into Eq. (2.28b), the following relationship between the components of T and n is determined

$$\frac{T_x}{l} = \frac{T_y}{m} = \frac{T_z}{n} \tag{2.28c}$$

These proportionalities indicate that the stress resultant must be parallel to the unit normal and therefore contains no shear component. Therefore, from Eqs. (2.24a), (2.24b) and (2.24c), we can write as below denoting the principal stress by σ_P

$$T_x = \sigma_P l \quad T_y = \sigma_P m \quad T_z = \sigma_P n \tag{2.28d}$$

These expressions together with Eqs. (2.24a), (2.24b) and (2.24c) lead to

$$\begin{aligned} (\sigma_x - \sigma_P)l + \tau_{xy} m + \tau_{xz} n &= 0 \\ \tau_{xy} l + (\sigma_y - \sigma_P)m + \tau_{yz} n &= 0 \\ \tau_{xy} l + \tau_{yz} m + (\sigma_z - \sigma_P)n &= 0 \end{aligned} \tag{2.29}$$

A non-trivial solution for the direction cosines requires that the characteristic determinant should vanish.

$$\begin{bmatrix} (\sigma_x - \sigma_P) & \tau_{xy} & \tau_{xz} \\ \tau_{xy} & (\sigma_y - \sigma_P) & \tau_{yz} \\ \tau_{xz} & \tau_{yz} & (\sigma_z - \sigma_P) \end{bmatrix} = 0 \tag{2.30}$$

Expanding (2.30) leads to

$$\sigma_P^3 - I_1 \sigma_P^2 + I_2 \sigma_P - I_3 = 0 \tag{2.31}$$

where

$$I_1 = \sigma_x + \sigma_y + \sigma_z \tag{2.31a}$$

$$I_2 = \sigma_x\sigma_y + \sigma_y\sigma_z + \sigma_z\sigma_x - \tau_{xy}^2 - \tau_{yz}^2 - \tau_{xz}^2 \tag{2.30b}$$

$$I_3 = \begin{vmatrix} \sigma_x & \tau_{xy} & \tau_{xz} \\ \tau_{xy} & \sigma_y & \tau_{yz} \\ \tau_{xz} & \tau_{yz} & \sigma_z \end{vmatrix} \tag{2.31c}$$

The three roots of Eq. (2.31) are the principal stresses, corresponding to which are three sets of direction cosines that establish the relationship of the principal planes to the origin of the non-principal axes.

2.15 Stress Invariant

Invariant means those quantities that are permanent, unexchangeable and do not vary under different conditions. In the context of stress tensor, invariants are such quantities that do not change with rotation of axes or which remain unaffected under transformation, from one set of axes to another. Therefore, the combination of stresses at a point that do not change with the orientation of co-ordinate axes is called stress invariants. Hence, the definition from Eq. (2.31)

$$\sigma_x + \sigma_y + \sigma_z = I_1 = \text{First invariant of stress}$$
$$\sigma_x\sigma_y + \sigma_y\sigma_z + \sigma_z\sigma_x - \tau_{xy}^2 - \tau_{yz}^2 - \tau_{zx}^2 = I_2 = \text{Second invariant of stress}$$
$$\sigma_x\sigma_y\sigma_z - \sigma_x\tau_{yz}^2 - \sigma_y\tau_{xz}^2 - \sigma_z\tau_{xy}^2 + 2\tau_{xy}\tau_{yz}\tau_{xz} = I_3 = \text{Third invariant of stress}$$

2.16 Equilibrium of a Two-Dimensional or Plane Element Differential Element

When a body is in equilibrium, any isolated part of the body is acted upon by an equilibrium set of forces. The small element with unit thickness shown in Fig. 2.11 represents part of a body and therefore must be in equilibrium if the entire body is to be in equilibrium.

It is to be noted that the components of stress generally vary from point to point in a stressed body. These variations are governed by the conditions of equilibrium

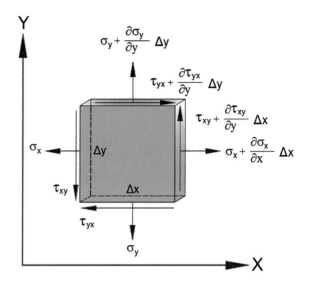

Fig. 2.11 Stress components acting on element

of statics. Fulfilment of these conditions establishes certain relationships, known as the differential equations of equilibrium. These involve the derivatives of the stress components.

Assume that $\sigma_x, \sigma_y, \tau_{xy}$ and τ_{yx} are functions of x and y but do not vary throughout the thickness (are independent of z) and that the other stress components are zero.

Also assume that the x and y components of the body forces per unit volume, F_x and F_y, are independent of z and that the z component of the body force $F_z = 0$. As the element is very small, the stress components may be considered to be distributed uniformly over each face.

Now, taking moments of force about the lower left corner and equating to zero,

$$- (\sigma_x \Delta y)\frac{\Delta y}{2} + (\tau_{xy}\Delta y)\frac{1}{2} - \left(\sigma_y + \frac{\partial \sigma_y}{\partial y}\Delta y\right)\Delta x\frac{\Delta x}{2} + \left(\tau_{yx} + \frac{\partial \tau_{yx}}{\partial y}\Delta y\right)\Delta x \Delta y$$

$$- \left(\tau_{xy} + \frac{\partial \tau_{xy}}{\partial x}\Delta x\right)\Delta x \Delta y + \left(\sigma_x + \frac{\partial \sigma_x}{\partial x}\Delta x\right)\Delta y\frac{\Delta y}{2} + \sigma_y \Delta x\frac{\Delta x}{2} - \tau_{yx}\Delta x\frac{1}{2}$$

$$+ (F_x \Delta y \Delta x)\frac{\Delta y}{2} - F_y \Delta x \Delta y\frac{\Delta x}{2} = 0$$

Neglecting the higher terms involving Δx and Δy and simplifying, the above expression is reduced to

$$\tau_{xy}\,\Delta x\,\Delta y = \tau_{yx}\,\Delta x\,\Delta y$$

or

$$\tau_{xy} = \tau_{yx}$$

In a like manner, it may be shown that

$$\tau_{yz} = \tau_{zy} \text{ and } \tau_{xz} = \tau_{zx}$$

Now, from the equilibrium of forces in x-direction, we obtain

$$-\sigma_x \, \Delta y + \left(\sigma_x + \frac{\partial \sigma_x}{\partial x} \Delta x\right) \Delta y + \left(\tau_{yx} + \frac{\partial \tau_{yx}}{\partial y} \Delta y\right) \Delta x - \tau_{yx} \Delta x + F_x \Delta x \Delta y = 0$$

Simplifying, we get

$$\frac{\partial \sigma_x}{\partial x} + \frac{\partial \tau_{yx}}{\partial y} + F_x = 0$$

or

$$\frac{\partial \sigma_x}{\partial x} + \frac{\partial \tau_{xy}}{\partial y} + F_x = 0$$

A similar expression is written to describe the equilibrium of y forces. The x and y equations yield the following differential equations of equilibrium.

$$\frac{\partial \sigma_x}{\partial x} + \frac{\partial \tau_{xy}}{\partial y} + F_x = 0 \tag{2.32a}$$

or

$$\frac{\partial \sigma_y}{\partial y} + \frac{\partial \tau_{xy}}{\partial x} + F_y = 0 \quad \text{since } \tau_{xy} = \tau_{yx} \tag{2.32b}$$

The differential equations of equilibrium for the case of three-dimensional stress may be generalized from the above expressions as follows (Fig. 2.12).

$$\frac{\partial \sigma_x}{\partial x} + \frac{\partial \tau_{xy}}{\partial y} + \frac{\partial \tau_{xz}}{\partial z} + F_x = 0 \tag{2.33a}$$

$$\frac{\partial \sigma_y}{\partial y} + \frac{\partial \tau_{xy}}{\partial x} + \frac{\partial \tau_{yz}}{\partial z} + F_y = 0 \tag{2.33b}$$

$$\frac{\partial \sigma_z}{\partial z} + \frac{\partial \tau_{xz}}{\partial x} + \frac{\partial \tau_{yz}}{\partial y} + F_z = 0 \tag{2.33c}$$

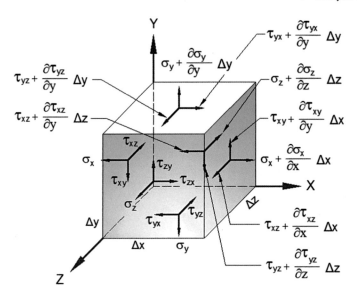

Fig. 2.12 Stress components acting on parallelopiped

2.17 Octahedral Stresses

A plane which is equally inclined to the three axes of reference is called the octahedral plane, and its direction cosines are $\ell = \pm\frac{1}{\sqrt{3}}, m = \pm\frac{1}{\sqrt{3}}, n = \pm\frac{1}{\sqrt{3}}$. The normal and shearing stresses acting on this plane are called the octahedral normal stress and octahedral shearing stress, respectively. In Fig. 2.13, x-, y- and z-axes are parallel to the principal axes and the octahedral planes defined with respect to the principal

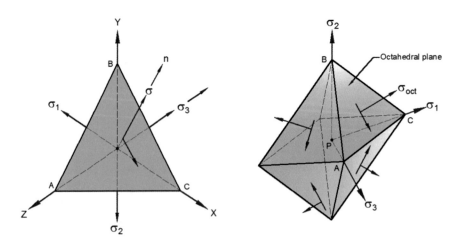

Fig. 2.13 Octahedral plane and octahedral stresses

axes and not with reference to an arbitrary frame of reference. Now, denoting the direction cosines of the plane ABC by l, m, and n, Eqs. (2.24a), (2.24b) and (2.24c) with $\sigma_x = \sigma_1$, $\tau_{xy} = \tau_{xz} = 0$, etc., reduce to

$$T_x = \sigma_1 l, \ T_y = \sigma_2 m \text{ and } T_z = \sigma_3 n \tag{2.34}$$

The resultant stress on the oblique plane is thus

$$T^2 = \sigma_1^2 l^2 + \sigma_2^2 m^2 + \sigma_3^2 n^2 = \sigma^2 + \tau^2$$
$$\therefore \ T^2 = \sigma^2 + \tau^2 \tag{2.35}$$

The normal stress on this plane is given by

$$\sigma = \sigma_1 l^2 + \sigma_2 m^2 + \sigma_3 n^2 \tag{2.36}$$

and the corresponding shear stress is

$$\tau = \left[(\sigma_1 - \sigma_2)^2 l^2 m^2 + (\sigma_2 - \sigma_3)^2 m^2 n^2 + (\sigma_3 - \sigma_1)^2 n^2 l^2\right]^{\frac{1}{2}} \tag{2.37}$$

The direction cosines of the octahedral plane are:

$$l = \pm\frac{1}{\sqrt{3}}, \quad m = \pm\frac{1}{\sqrt{3}}, \quad n = \pm\frac{1}{\sqrt{3}}$$

Substituting in (2.35), (2.36), (2.37), we get

$$\text{Resultant stress } T = \sqrt{\frac{1}{3}(\sigma_1^2 + \sigma_2^2 + \sigma_3^2)} \tag{2.38}$$

$$\text{Normal stress} = \sigma = \frac{1}{3}(\sigma_1 + \sigma_2 + \sigma_3) \tag{2.39}$$

$$\text{Shear stress} = \tau = \frac{1}{3}\sqrt{(\sigma_1 - \sigma_2)^2 + (\sigma_2 - \sigma_3)^2 + (\sigma_3 - \sigma_1)^2} \tag{2.40}$$

Also,

$$\tau = \frac{1}{3}\sqrt{2(\sigma_1 + \sigma_2 + \sigma_3)^2 - 6(\sigma_1\sigma_2 + \sigma_2\sigma_3 + \sigma_1\sigma_3)} \tag{2.41}$$

$$\tau = \frac{1}{3}\sqrt{2I_1^2 - 6I_2} \tag{2.42}$$

2.18   Mohr's Stress Circle

A graphical means of representing the stress relationships was discovered by
Culmann (1866) and developed in detail by Mohr (1882), after whom the graphical
method is now named.

2.19   Mohr Circles for Two-Dimensional Stress Systems

Biaxial Compression (Fig. 2.14a)
The biaxial stresses are represented by a circle that plots in positive σ space, passing
through stress points σ_1, σ_2 on the $\tau = 0$ axis. The centre of the circle is located on
the $\tau = 0$ axis at stress point $\frac{1}{2}(\sigma_1 + \sigma_2)$. The radius of the circle has the magnitude
$\frac{1}{2}(\sigma_1 - \sigma_2)$, which is equal to τ_{max}.

Biaxial Compression/Tension (Fig. 2.14b)
Here, the stress circle extends into both positive and negative σ space. The centre
of the circle is located on the $\tau = 0$ axis at stress point $\frac{1}{2}(\sigma_1 + \sigma_2)$ and has radius
$\frac{1}{2}(\sigma_1 - \sigma_2)$. This is also the maximum value of shear stress, which occurs in a direction at 45° to the σ_1 direction. The normal stress is zero in directions $\pm \theta$ to the
direction of σ_1, where

$$\cos 2\theta = -\frac{\sigma_1 + \sigma_2}{\sigma_1 - \sigma_2}$$

Biaxial Pure Shear (Fig. 2.14c)
Here, the circle has a radius equal to τ_{zy}, which is equal in magnitude to τ_{yz}, but
opposite in sign. The centre of circle is at $\sigma = 0$, $\tau = 0$. The principal stresses σ_1,
σ_2 are equal in magnitude, but opposite in sign, and are equal in magnitude to τ_{zy}.
The directions of σ_1, σ_2 are at 45° to the directions of τ_{zy}, τ_{yz}.

2.20   Construction of Mohr's Circle for Two- Dimensional
Stress System

Sign Convention
For the purposes of constructing and reading values of stress from Mohr's circle, the
sign convention for shear stress is as follows.

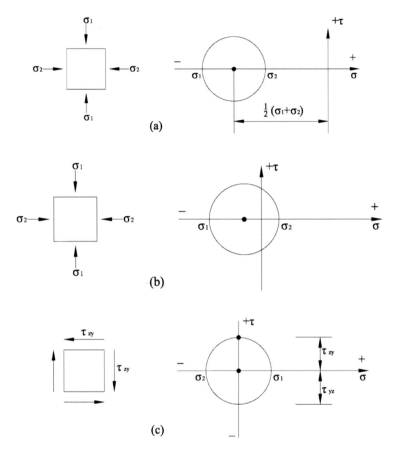

Fig. 2.14 Simple biaxial stress systems: **a** compression, **b** tension/compression, **c** pure shear

If the shearing stresses on opposite faces of an element would produce shearing forces that result in a clockwise couple, these stresses are regarded as "positive".

Procedure for Obtaining Mohr's Circle

(1) Establish a rectangular co-ordinate system, indicating $+\tau$ and $+\sigma$. Both stress scales must be identical.

(2) Locate the centre C of the circle on the horizontal axis a distance $\frac{1}{2}(\sigma_X + \sigma_Y)$ from the origin as shown in the figure above.

(3) Locate point A by co-ordinates $\sigma_x, -\tau_{xy}$.

(4) Locate the point B by co-ordinates σ_y, τ_{xy}.

(5) Draw a circle with centre C and of radius equal to CA.

(6) Draw a line AB through C.

An angle of 2θ on the circle corresponds to an angle of θ on the element. The state of stress associated with the original x and y planes corresponds to points A and

B on the circle, respectively. Points lying on the diameter other than *AB*, such as *A'* and *B'*, define state of stress with respect to any other set of *x'* and *y'* planes rotated relative to the original set through an angle θ (Fig. 2.15).

It is clear from the figure that the points A_1 and B_1 on the circle locate the principal stresses and provide their magnitudes as defined by Eqs. (2.14) and (2.15), while *D* and *E* represent the maximum shearing stresses. The maximum value of shear stress (regardless of algebraic sign) will be denoted by τ_{max} and are given by

$$\tau_{max} = \pm\frac{1}{2}(\sigma_1 - \sigma_2) = \pm\sqrt{\left(\frac{\sigma_x - \sigma_y}{2}\right)^2 + \tau_{xy}^2} \qquad (2.43)$$

Mohr's circle shows that the planes of maximum shear are always located at 45° from planes of principal stress.

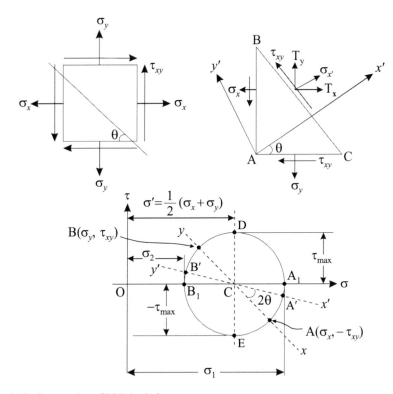

Fig. 2.15 Construction of Mohr's circle

2.21 Equilibrium Equations in Polar Co-ordinates (Two-Dimensional State of Stress)

While discussing the problems with circular boundaries, it is more convenient to use the cylindrical co-ordinates such as r, θ and z. In the case of plane stress or plane strain problems, we have $\tau_{rz} = \tau_{\theta z} = 0$ and the other stress components as functions of r and θ only. Hence, the cylindrical co-ordinates reduce to polar co-ordinates in this case. In general, polar co-ordinates are used advantageously where a degree of axial symmetry exists. Examples include a cylinder, a disc, a curved beam and a large thin plate containing a circular hole.

The polar co-ordinate system (r, θ) and the cartesian system (x, y) are related by the following expressions:

$$x = r\cos\theta, \quad r^2 = x^2 + y^2$$
$$y = r\sin\theta, \quad \theta = \tan^{-1}\left(\tfrac{y}{x}\right) \tag{2.44}$$

Consider the state of stress on an infinitesimal element $a\,b\,c\,d$ of unit thickness described by the polar co-ordinates as shown in Fig. 2.16. The body forces denoted by F_r and F_θ are directed along r and θ directions, respectively.

Resolving the forces in the r-direction, we have for equilibrium, $\Sigma F_r = 0$,

$$-\sigma_r \times r\,d\theta + \left(\sigma_r + \frac{\partial \sigma_r}{\partial r}dr\right)(r + dr)d\theta - \sigma_\theta dr \sin\frac{d\theta}{2}$$
$$+ F_r - \left(\sigma_\theta + \frac{\partial \sigma_\theta}{\partial \theta}d\theta\right)dr\sin\frac{d\theta}{2} - \tau_{r\theta}dr\cos\frac{d\theta}{2} + \left(\tau_{r\theta} + \frac{\partial \tau_{r\theta}}{\partial \theta}d\theta\right)dr\cos\frac{d\theta}{2} = 0$$

Since $d\theta$ is very small,

Fig. 2.16 Stresses acting on an element

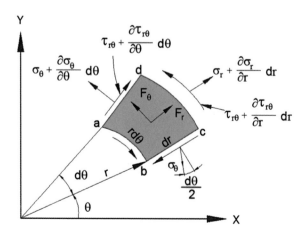

$$\sin \frac{d\theta}{2} = \frac{d\theta}{2} \text{ and } \cos \frac{d\theta}{2} = 1$$

Neglecting higher order terms and simplifying, we get

$$r \frac{\partial \sigma_r}{\partial r} dr \, d\theta + \sigma_r dr \, d\theta - \sigma_\theta dr \, d\theta + \frac{\partial \tau_{r\theta}}{\partial \theta} dr \, d\theta = 0$$

on dividing throughout by $rd\theta \, dr$, we have

$$\frac{\partial \sigma_r}{\partial r} + \frac{1}{r} \frac{\partial \tau_{r\theta}}{\partial \theta} + \frac{\sigma_r - \sigma_\theta}{r} + F_r = 0 \tag{2.45}$$

Similarly resolving all the forces in the θ-direction at right angles to r-direction, we have

$$- \sigma_\theta dr \cos \frac{d\theta}{2} + \left(\sigma_\theta + \frac{\partial \sigma_\theta}{\partial \theta} d\theta \right) dr \, \cos \frac{d\theta}{2} + \tau_{r\theta} dr \sin \frac{d\theta}{2} + \left(\tau_{r\theta} + \frac{\partial \tau_{r\theta}}{\partial \theta} d\theta \right) dr$$

$$\sin \frac{d\theta}{2} - \tau_{r\theta} r d\theta + (r + dr) \, d\theta \left(\tau_{r\theta} + \frac{\partial \tau_{r\theta}}{\partial r} dr \right) + F_\theta = 0$$

On simplification, we get

$$\left(\frac{\partial \sigma_\theta}{\partial \theta} + \tau_{r\theta} + \tau_{r\theta} + r \frac{\partial \tau_{r\theta}}{\partial r} \right) d\theta \, dr = 0$$

Dividing throughout by $rd\theta \, dr$, we get

$$\frac{1}{r} \cdot \frac{\partial \sigma_\theta}{\partial \theta} + \frac{\partial \tau_{r\theta}}{\partial r} + \frac{2\tau_{r\theta}}{r} + F_\theta = 0 \tag{2.46}$$

In the absence of body forces, the equilibrium equations can be represented as:

$$\frac{\partial \sigma_r}{\partial r} + \frac{1}{r} \frac{\partial \tau_{r\theta}}{\partial \theta} + \frac{\sigma_r - \sigma_\theta}{r} = 0$$

$$\frac{1}{r} \frac{\partial \sigma_\theta}{\partial \theta} + \frac{\partial \tau_{r\theta}}{\partial r} + \frac{2\tau_{r\theta}}{r} = 0 \tag{2.47}$$

2.22   General State of Stress in Three Dimensions in Cylindrical Co-ordinate System

See Fig. 2.17.

The equilibrium equations for three-dimensional state are given by

Fig. 2.17 Stresses acting on the three-dimensional element

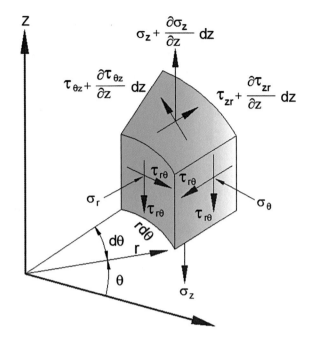

$$\frac{\partial \sigma_r}{\partial r} + \frac{1}{r}\frac{\partial \tau_{r\theta}}{\partial \theta} + \frac{\partial \tau_{zr}}{\partial z}\left(\frac{\sigma_r - \sigma_\theta}{r}\right) = 0 \tag{2.48}$$

$$\frac{\partial \tau_{r\theta}}{\partial r} + \frac{1}{r}\frac{\partial \sigma_\theta}{\partial \theta} + \frac{\partial \tau_{\theta z}}{\partial z}\frac{2\tau_{r\theta}}{r} = 0 \tag{2.49}$$

$$\frac{\partial \tau_{zr}}{\partial r} + \frac{1}{r}\frac{\partial \tau_{\theta z}}{\partial \theta} + \frac{\partial \sigma_z}{\partial z} + \frac{\tau_{zr}}{r} = 0 \tag{2.50}$$

2.23 Numerical Examples

Example 2.1 When the stress tensor at a point with reference to axes (x, y, z) is given by the array,

$$\begin{bmatrix} 4 & 1 & 2 \\ 1 & 6 & 0 \\ 2 & 0 & 8 \end{bmatrix} \text{ MPa}$$

show that by transformation of the axes by 45° about the z-axis, the stress invariants remain unchanged.

Solution The stress invariants are

$$I_1 = 4 + 6 + 8 = 18$$
$$I_2 = 4 \times 6 + 6 \times 8 + 4 \times 8 - 1 \times 1 - 2 \times 2 - 0 = 99$$
$$I_3 = 4 \times 48 - 1 \times 8 + 2 \times (-12) = 160$$

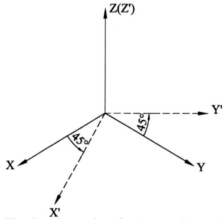

The direction cosines for the transformation are given by

	x	y	z
x'	$\frac{1}{\sqrt{2}}$	$\frac{1}{\sqrt{2}}$	0
y'	$-\frac{1}{\sqrt{2}}$	$\frac{1}{\sqrt{2}}$	0
z'	0	0	1

Using Eqs. (2.27), (2.27a), (2.27b), (2.27c), (2.27d) and (2.27e), we get

$$\sigma_{x'} = 4 \times \frac{1}{2} + 6 \times \frac{1}{2} + 0 + 2 \times 1 \times \frac{1}{2} + 0 + 0$$
$$= 6 \text{ MPa}$$

$$\sigma_{y'} = 4 \times \frac{1}{2} + 6 \times \frac{1}{2} + 0 - 2 \times 1 \times \frac{1}{2} + 0 + 0$$
$$= 4 \text{ MPa}$$

$$\sigma_{z'} = 0 + 0 + 8 \times 1 + 0 + 0 + 0$$
$$= 8 \text{ MPa}$$

$$\tau_{x'y'} = -4 \times \frac{1}{2} + 6 \times \frac{1}{2} + 0 + 1\left(\frac{1}{2} - \frac{1}{2}\right) + 0 + 0$$

$$= 1 \text{ MPa}$$

$$\tau_{y'z'} = 0 + 0 + 0 + 0 + 0 + 2\left(-\frac{1}{\sqrt{2}}\right)$$

$$= -\sqrt{2} \text{ MPa}$$

$$\tau_{x'z'} = 0 + 0 + 0 + 0 + 0 + 2\left(\frac{1}{\sqrt{2}}\right)$$

$$= \sqrt{2} \text{ MPa}$$

Hence, the new stress tensor becomes

$$\begin{bmatrix} 6 & 1 & \sqrt{2} \\ 1 & 4 & -\sqrt{2} \\ \sqrt{2} & -\sqrt{2} & 8 \end{bmatrix} \text{ MPa}$$

Now, the new invariants are

$$I_1' = 6 + 4 + 8 = 18$$
$$I_2' = 6 \times 4 + 4 \times 8 + 6 \times 8 - 1 - 2 - 2 = 99$$
$$I_3' = 6 \times 30 - 1 \times 10 + \sqrt{2}\left(-\frac{5}{\sqrt{2}}\right) = 160$$

which remains unchanged. Hence proved.

Example 2.2 The state of stress at a point is given by the following array of terms

$$\begin{bmatrix} 9 & 6 & 3 \\ 6 & 5 & 2 \\ 3 & 2 & 4 \end{bmatrix} \text{ MPa}$$

Determine the principal stresses and principal directions.

Solution The principal stresses are the roots of the cubic equation

$$\sigma^3 - I_1\sigma^2 + I_2\sigma - I_3 = 0$$

Here,

$$I_1 = 9 + 5 + 4 = 18$$
$$I_2 = 9 \times 5 + 5 \times 4 + 9 \times 4 - (6)^2 - (2)^2 - (3)^2 = 52$$
$$I_3 = 9 \times 5 \times 4 - 9 \times 4 - 5 \times 9 - 4 \times 36 + 2 \times 6 \times 2 \times 3 = 27$$

∴ The cubic equation becomes

$$\sigma^3 - 18\sigma^2 + 52\sigma - 27 = 0$$

The roots of the cubic equation are the principal stresses. Hence, the three principal stresses are

$$\sigma_1 = 14.554 \text{ MPa}; \ \sigma_2 = 2.776 \text{ MPa and } \sigma_3 = 0.669 \text{ MPa}$$

Now to find principal directions for σ_1 stress:

$$\begin{vmatrix} (9 - 14.554) & 6 & 3 \\ 6 & (5 - 14.554) & 2 \\ 3 & 2 & (4 - 14.554) \end{vmatrix}$$

$$= \begin{vmatrix} -5.554 & 6 & 3 \\ 6 & -9.554 & 2 \\ 3 & 2 & -10.554 \end{vmatrix}$$

$$A = \begin{bmatrix} -9.554 & 2 \\ 2 & -10.554 \end{bmatrix} = 100.83 - 4 = 96.83$$

$$B = -\begin{bmatrix} 6 & 2 \\ 3 & -10.554 \end{bmatrix} = -(-63.324 - 6) = 69.324$$

$$C = \begin{bmatrix} 6 & -9.554 \\ 3 & 2 \end{bmatrix} = 12 + 28.662 = 40.662$$

$$\sqrt{A^2 + B^2 + C^2}$$

$$= \sqrt{(96.83)^2 + (69.324)^2 + (40.662)^2}$$

$$= 125.83$$

$$l_1 = \frac{A}{\sqrt{A^2 + B^2 + C^2}} = \frac{96.53}{125.83} = 0.769$$

$$m_1 = \frac{B}{\sqrt{A^2 + B^2 + C^2}} = \frac{69.324}{125.83} = 0.550$$

$$n_1 = \frac{C}{\sqrt{A^2 + B^2 + C^2}} = \frac{40.662}{125.84} = 0.325$$

Similarly, the principal stress directions for σ_2 stress and σ_3 stress are calculated. Therefore,

$$l_2 = -0.226 \quad l_3 = 0.596$$
$$m_2 = -0.177 \quad m_3 = -0.800$$
$$n_2 = 0.944 \quad n_3 = 0.057$$

Fig. 2.18 Stresses in structural member

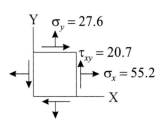

Example 2.3 At a point in the structural member, the stresses (in MPa) are represented as in Fig. 2.18. Employ Mohr's circle to determine:

(a) Magnitude and orientation of the principal stresses
(b) Magnitude and orientation of the maximum shearing stresses and associated normal stresses.

In each case show the results on a properly oriented element.

Solution Centre of the Mohr's circle $= OC$

$$= \frac{27.6 + 55.2}{2} = 41.4 \text{ MPa}$$

(a) Principal stresses are represented by points A_1 and B_1. Hence, the maximum and minimum principal stresses, referring to the circle, are

$$\sigma_{1,2} = 41.4 \pm \sqrt{\tfrac{1}{4}(55.2 - 27.6)^2 + (20.7)^2}$$
$$\sigma_1 = 66.3 \text{ MPa and } \sigma_2 = 16.5 \text{ MPa}$$

The planes on which the principal stresses act are given by

$$2\theta'_p = \tan^{-1} \frac{20.7}{13.8} = 56.30°$$

and

$$2\theta''_p = 56.30 + 180 = 236.30°$$

Hence,

$$\theta'_p = 28.15° \text{ and } \theta''_p = 118.15°$$

Mohr's circle clearly indicates that θ'_p locates the σ_1 plane.

(b) The maximum shearing stresses are given by points D and E. Thus,

$$\tau_{max} = \pm\sqrt{\frac{1}{4}(55.2 - 27.6)^2 + (20.7)^2}$$

$$= \pm24.9 \text{ MPa}$$

The planes on which these stresses act are represented by

$$\theta'_s = 28.15° + 45° = 73.15°$$

and

$$\theta''_s = 163.15°$$

See Fig. 2.19.

Example 2.4 The stress (in N/m^2) acting on an element of a loaded body is shown in Fig. 2.20. Apply Mohr's circle to determine the normal and shear stresses acting on a plane defined by $\theta = 30°$.

Solution The Mohr's circle drawn below describes the state of stress for the given element. Points A_1 and B_1 represent the stress components on the x- and y-faces, respectively. The radius of the circle is $(14 + 28)\frac{10^6}{2} = 21 \times 10^6$. Corresponding to the 30° plane within the element, it is necessary to rotate through 60° counter

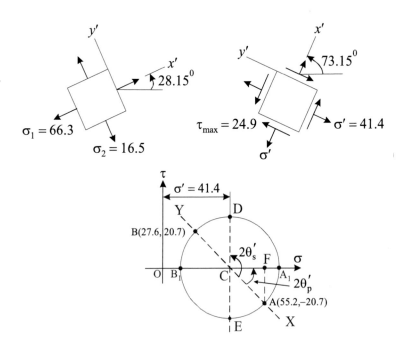

Fig. 2.19 Mohr's stress circle

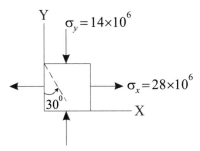

Fig. 2.20 Stresses in element of a body

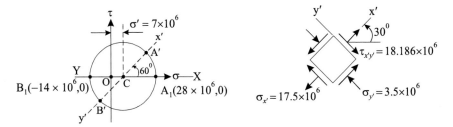

Fig. 2.21 Mohr's stress circle

clockwise on the circle to locate point A'. A 240° counterclockwise rotation locates point B' (Fig. 2.21).

From the above Mohr's circle,

$$\sigma_{x'} = (7 + 21 \cos 60°)10^6 = 17.5 \times 10^6 \text{ N/m}^2$$
$$\sigma_{y'} = -3.5 \times 10^6 \text{ N/m}^2$$
$$\tau_{x'y'} = \pm 21 \times 10^6 \sin 60° = \pm 18.86 \times 10^6 \text{ N/m}^2$$

Example 2.5 A rectangular bar of metal of cross section 30 mm × 25 mm is subjected to an axial tensile force of 180 KN. Calculate the normal, shear and resultant stresses on a plane whose normal has the following direction cosines:

(i) $l = m = \dfrac{1}{\sqrt{2}}$ and $n = 0$

(ii) $l = m = n = \dfrac{1}{\sqrt{3}}$

Solution Let normal stress acting on the cross section is given by σ_y.

$$\therefore \quad \sigma_y = \frac{\text{Axial load}}{\text{cross} - \text{sectional area}}$$

$$= \frac{180 \times 10^3}{30 \times 25}$$

$$= 240 \, \text{N/mm}^2$$

Now, by Cauchy's formula, the stress components along x, y and z co-ordinates are

$$T_x = \sigma_x l + \tau_{xy} m + \tau_{xz} n$$
$$T_y = \tau_{xy} l + \sigma_y m + \tau_{yz} n$$
$$T_z = \tau_{xz} l + \tau_{yz} m + \sigma_z n \qquad \text{(a)}$$

And the normal stress acting on the plane whose normal has the direction cosines l, m and n is,

$$\sigma = T_x l + T_y m + T_z n \qquad \text{(b)}$$

Case (i) For $l = m = \frac{1}{\sqrt{2}}$ and $n = 0$

Here,

$$\sigma_x = 0, \ \tau_{xy} = 0, \ \sigma_y = 240 \, \text{N/mm}^2$$

$$\tau_{xz} = 0, \ \tau_{yz} = 0, \ \sigma_z = 0$$

Substituting the above in (a), we get

$$T_x = 0, \ T_y = \sigma_y m = \frac{240}{\sqrt{2}}, \ T_z = 0$$

Substituting in (b), we get

$$\sigma = 0 + \frac{240}{\sqrt{2}} \left(\frac{1}{\sqrt{2}} \right) + 0 = 120 \, \text{N/mm}^2$$

Resultant stress on the plane is

$$T = \sqrt{T_x^2 + T_y^2 + T_z^2}$$

$$= \sqrt{0 + \left[\frac{240}{\sqrt{2}} \right]^2 + 0}$$

$$T = 169.706 \, \text{N/mm}^2$$

But shear stress τ can be determined from the relation

$$T^2 = \sigma^2 + \tau^2$$

or

$$\tau = \sqrt{T^2 - \sigma^2}$$
$$= \sqrt{(169.706)^2 - (120)^2}$$
$$\tau = 120 \text{ N/mm}^2$$

Case (ii) For $l = m = n = \frac{1}{\sqrt{3}}$

Again from (a),

$$T_x = 0, T_y = \sigma_y m = \frac{240}{\sqrt{3}}, T_z = 0$$

Normal stress $= \sigma = 0 + \frac{240}{\sqrt{3}}\left(\frac{1}{\sqrt{3}}\right) + 0 = 80.00 \text{ N/mm}^2$

Resultant stress on the plane is

$$T = \sqrt{T_x^2 + T_y^2 + T_z^2}$$
$$T = \sqrt{0 + \left[\frac{240}{\sqrt{3}}\right]^2 + 0}$$
$$\tau = 113.13 \text{ N/mm}^2$$
$$\text{Shear stress } = \tau = \sqrt{(138.56)^2 - (80)^2}$$
$$\tau = 113.13 \text{ N/mm}^2$$

Example 2.6 A body is subjected to three-dimensional forces and the state of stress at a point in it is represented as

$$\begin{bmatrix} 200 & 200 & 200 \\ 200 & -100 & 200 \\ 200 & 200 & -100 \end{bmatrix} \text{MPa}$$

Determine the normal stress, shearing stress and resultant stress on the octahedral plane.

Solution For the octahedral plane, the direction cosines are

$$l = m = n = \frac{1}{\sqrt{3}}$$

Here,

$$\sigma_x = 200 \text{ MPa}$$
$$\sigma_y = -100 \text{ MPa}$$
$$\sigma_y = -100 \text{ MPa}$$
$$\tau_{xy} = \tau_{yz} = \tau_{zx} = 200 \text{ MPa}$$

Substituting the above in Cauchy's formula, we get

$$T_x = 200\left(\frac{1}{\sqrt{3}}\right) + 200\left(\frac{1}{\sqrt{3}}\right) + 200\left(\frac{1}{\sqrt{3}}\right) = 346.41 \text{ MPa}$$

$$T_y = 200\left(\frac{1}{\sqrt{3}}\right) - 100\left(\frac{1}{\sqrt{3}}\right) + 200\left(\frac{1}{\sqrt{3}}\right) = 173.20 \text{ MPa}$$

$$T_z = 200\left(\frac{1}{\sqrt{3}}\right) + 200\left(\frac{1}{\sqrt{3}}\right) - 100\left(\frac{1}{\sqrt{3}}\right) = 173.20 \text{ MPa}$$

Normal stress on the plane is given by

$$\sigma = T_x l + T_y m + T_z n$$
$$= 346.41\left(\frac{1}{\sqrt{3}}\right) + 173.20\left(\frac{1}{\sqrt{3}}\right) + 173.20\left(\frac{1}{\sqrt{3}}\right)$$
$$\sigma = 400 \text{ MPa}$$

$$\text{Resultant stress} = T = \sqrt{T_x^2 + T_y^2 + T_z^2}$$
$$= \sqrt{(346.41)^2 + (173.20)^2 + (173.20)^2}$$
$$T = 424.26 \text{ MPa}$$
$$\text{Also, tangential stress} = \tau = \sqrt{(424.26)^2 - (400)^2}$$
$$= 141.41 \text{ MPa}$$

Also,

$$\text{tangential stress} = \tau = \sqrt{(424.26)^2 - (400)^2}$$
$$= 141.41 \text{ MPa}$$

Example 2.7 The state of stress at a point is given as follows:

$$\sigma_x = -800 \text{ kPa}, \quad \sigma_y = 1200 \text{ kPa}, \quad \sigma_z = -400 \text{ kPa}$$
$$\tau_{xy} = 400 \text{ kPa}, \quad \tau_{yz} = -600 \text{ kPa}, \quad \tau_{zx} = 500 \text{ kPa}$$

Determine (a) the stresses on a plane whose normal has direction cosines $l = \frac{1}{4}$, $m = \frac{1}{2}$ and (b) the normal and shearing stresses on that plane.

Solution We have the relation,

$$l^2 + m^2 + n^2 = 1$$

$$\therefore \quad \left(\frac{1}{4}\right)^2 + \left(\frac{1}{2}\right)^2 + n^2 = 1$$

$$\therefore \quad n = \frac{\sqrt{11}}{4}$$

(a) Using Cauchy's formula,

$$T_x = -800\left(\frac{1}{4}\right) + 400\left(\frac{1}{2}\right) + 500\left(\frac{\sqrt{11}}{4}\right) = 414.60 \text{ kPa}$$

$$T_y = 400\left(\frac{1}{4}\right) + 1200\left(\frac{1}{2}\right) - 600\left(\frac{\sqrt{11}}{4}\right) = 202.51 \text{ kPa}$$

$$T_z = 500\left(\frac{1}{4}\right) - 600\left(\frac{1}{2}\right) - 400\left(\frac{\sqrt{11}}{4}\right) = -506.66 \text{ kPa}$$

(b) **Normal stress,**

$$\sigma = T_x l + T_y m + T_z n$$

$$= 414.60\left(\frac{1}{4}\right) + 202.51\left(\frac{1}{2}\right) - 506.66\left(\frac{\sqrt{11}}{4}\right)$$

$$\sigma = -215.20 \text{ kPa}$$

$$\text{Resultant stress on the plane} = T = \sqrt{(41460)^2 + (20251)^2 + (50666)^2}$$
$$= 685.28 \text{ MPa}$$

$$\text{Shear stress on the plane} = \tau = \sqrt{(685.28)^2 - (-215.20)^2}$$
$$= 650.61 \text{ kPa}$$

Example 2.8 Given the state of stress at a point as below

$$\begin{bmatrix} 100 & 80 & 0 \\ 90 & -60 & 0 \\ 0 & 0 & 40 \end{bmatrix} \text{ kPa}$$

Considering another set of co-ordinate axes, $x'\, y'\, z'$ in which z' coincides with z and x' is rotated by 30° anticlockwise from x-axis, determine the stress components in the new co-ordinates system.

Solution The direction cosines for the transformation are given by

	x	y	z
x'	0.866	0.5	0
y'	−0.5	0.866	0
z'	0	0	1

See Fig. 2.22.

Now using Eqs. (2.27), (2.27a)–(2.21e), we get

$$\sigma_{x^1} = 100(0.866)^2 - 60(0.5)^2 + 0 + 2[80 \times 0.866 \times 0.5 + 0 + 0]$$
$$\sigma_{x'} = 129.3\,\text{kPa}$$
$$\sigma_{y'} = 100(-0.5)^2 - 60(0.866)^2 + 0 + 2[80(-0.5)(0.866) + 0 + 0]$$
$$\sigma_{y'} = -89.3\,\text{kPa}$$
$$\sigma_{z'} = 0 + 0 + 40(1)^2 + 2[0 + 0 + 0]$$
$$\sigma_{z'} = 40\,\text{kPa}$$

$$\tau_{x'y'} = 100(0.866)(-0.5) - 60(0.5)(0.866) + 0$$
$$\qquad + 80[(0.866 \times 0.866) + (-0.5)(0.5)] + 0 + 0$$
$$\tau_{x'y'} = -29.3\,\text{kPa}$$
$$\tau_{y'z'} = 0 \text{ and } \tau_{z'x'} = 0$$

Therefore, the state of stress in new co-ordinate system is

Fig. 2.22 Co-ordinate system

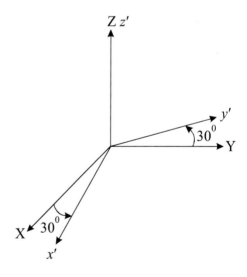

$$\begin{bmatrix} 129.3 & -29.3 & 0 \\ -29.3 & -89.3 & 0 \\ 0 & 0 & 40 \end{bmatrix} \text{(KPa)}$$

Example 2.9 The stress tensor at a point is given by the following array

$$\begin{bmatrix} 50 & -20 & 40 \\ -20 & 20 & 10 \\ 40 & 10 & 30 \end{bmatrix} \text{(kPa)}$$

Determine the stress vectors on the plane whose unit normal has direction cosines $\frac{1}{\sqrt{2}}, \frac{1}{2}, \frac{1}{2}$

Solution The stress vectors are given by

$$T_x = \sigma_x l + \tau_{xy} m + \tau_{xz} n \qquad\qquad (a)$$

$$T_y = \tau_{xy} l + \sigma_y m + \tau_{yz} n \qquad\qquad (b)$$

$$T_z = \tau_{xz} l + \tau_{yz} m + \sigma_z n \qquad\qquad (c)$$

Substituting the stress components in (a), (b) and (c), we get

$$T_x = 50\left(\frac{1}{\sqrt{2}}\right) - 20\left(\frac{1}{2}\right) + 40\left(\frac{1}{2}\right) = 45.35 \text{ kPa}$$

$$T_y = -20\left(\frac{1}{\sqrt{2}}\right) + 20\left(\frac{1}{2}\right) + 10\left(\frac{1}{2}\right) = 0.858 \text{ kPa}$$

$$T_z = 40\left(\frac{1}{\sqrt{2}}\right) + 10\left(\frac{1}{2}\right) + 30\left(\frac{1}{2}\right) = 48.28 \text{ kPa}$$

Now, resultant stress is given by

$$T = \left(45.35\,\hat{i} + 0.858\,\hat{j} + 48.28\hat{k}\right) \text{ kPa}$$

Example 2.10 The stress tensor at a point is given by the following array

$$\begin{bmatrix} 40 & 20 & 30 \\ 20 & 30 & 40 \\ 30 & 40 & 20 \end{bmatrix} \text{(kPa)}$$

Calculate the deviator and spherical stress tensors.

Solution

$$\text{Mean stress} = \sigma_m = \frac{1}{3}\left(\sigma_x + \sigma_y + \sigma_z\right)$$

$$= \frac{1}{3}(40 + 30 + 20)$$

$$= 30\,kPa$$

$$\text{Deviator stress tensor} = \begin{bmatrix} (\sigma_x - \sigma_m) & \tau_{xy} & \tau_{xz} \\ \tau_{xy} & \sigma_y - \sigma_m & \tau_{yz} \\ \tau_{xz} & \tau_{yz} & (\sigma_z - \sigma_m) \end{bmatrix}$$

$$= \begin{bmatrix} (40 - 30) & 20 & 30 \\ 20 & (30 - 30) & 40 \\ 30 & 40 & (20 - 30) \end{bmatrix}$$

$$= \begin{bmatrix} 10 & 20 & 30 \\ 20 & 0 & 40 \\ 30 & 40 & -10 \end{bmatrix} kPa$$

$$\text{Spherical stress tensor} = \begin{bmatrix} \sigma_m & 0 & 0 \\ 0 & \sigma_m & 0 \\ 0 & 0 & \sigma_m \end{bmatrix}$$

$$= \begin{bmatrix} 30 & 0 & 0 \\ 0 & 30 & 0 \\ 0 & 0 & 30 \end{bmatrix} kPa$$

Example 2.11 The stress components at a point in a body are given by

$$\sigma_x = 3xy^2z + 2x, \ \tau_{xy} = 0$$
$$\sigma_y = 5xyz + 3y \quad \tau_{yz} = \tau_{xz} = 3xy^2z + 2xy$$
$$\sigma_z = x^2y + y^2z$$

Determine whether these components of stress satisfy the equilibrium equations or not as the point $(1, -1, 2)$. If not, then determine the suitable body force required at this point so that these stress components become under equilibrium.

Solution The equations of equilibrium are given by

$$\frac{\partial \sigma_x}{\partial x} + \frac{\partial \tau_{xy}}{\partial y} + \frac{\partial \tau_{xz}}{\partial z} = 0 \tag{a}$$

$$\frac{\partial \tau_{xy}}{\partial x} + \frac{\partial \sigma_y}{\partial y} + \frac{\partial \tau_{yz}}{\partial z} = 0 \tag{b}$$

$$\frac{\partial \tau_{xz}}{\partial x} + \frac{\partial \tau_{yz}}{\partial y} + \frac{\partial \sigma_z}{\partial z} = 0 \tag{c}$$

Differentiating the stress components with respective axes, we get

$$\frac{\partial \sigma_x}{\partial x} = 3y^2 z + 2, \quad \frac{\partial \tau_{xy}}{\partial y} = 0, \quad \frac{\partial \tau_{xz}}{\partial z} = 3xy^2$$

Substituting in (a), $3y^2 z + 2 + 0 + 3xy^2$

At point $(1, -1, 2)$, we get $3 \times 1 \times 2 + 2 + 3 \times 1 \times 1 = 11$ which is not equal to zero.

Similarly,

$$\frac{\partial \sigma_y}{\partial y} = 5xz + 3, \quad \frac{\partial \tau_{yz}}{\partial z} = 3xy^2 + 0$$

$\therefore$ (ii) becomes $0 + 5xz + 3 + 3xy^2$

At point $(1, -1, 2)$, we get $5 \times 1 \times 2 + 3 + 3 \times 1 \times 1 = 16$ which is not equal to zero.

And

$$\frac{\partial \sigma_z}{\partial z} = y^2, \quad \frac{\partial \tau_{yz}}{\partial y} = 6xyz + 2x, \quad \frac{\partial \tau_{xz}}{\partial x} = 3y^2 z + 2y$$

Therefore, (iii) becomes $3y^2 z + 2y + 6xyz + 2x + y^2$

At the point $(1, -1, 2)$, we get $3 \times 1 \times 2 + 2 \times (-1) + 6 \times 1 \times (-1) \times 2 + 2 \times 1 + (-1)^2 = -5$ which is not equal to zero.

Hence, the given stress components do not satisfy the equilibrium equations.

Recalling (a), (b) and (c) with body forces, the equations can be modified as below.

$$\frac{\partial \sigma_x}{\partial x} + \frac{\partial \tau_{xy}}{\partial y} + \frac{\partial \tau_{xz}}{\partial z} + F_x = 0 \tag{d}$$

$$\frac{\partial \tau_{xy}}{\partial x} + \frac{\partial \sigma_y}{\partial y} + \frac{\partial \tau_{yz}}{\partial z} + F_y = 0 \tag{e}$$

$$\frac{\partial \tau_{xz}}{\partial x} + \frac{\partial \tau_{yz}}{\partial y} + \frac{\partial \sigma_z}{\partial z} + F_z = 0 \tag{f}$$

where F_x, F_y and F_z are the body forces.

Substituting the values in (d), (e) and (f), we get body forces so that the stress components become under equilibrium.

Therefore,

$$3 \times 1 \times 2 + 2 + 3 \times 1 \times 1 + F_x = 0$$
$$\therefore \quad F_x = -11$$

Also,

$$5 \times 1 \times 2 + 3 + 3 \times 1 \times 1 + F_y = 0$$

$$\therefore F_y = -16$$

and

$$3 \times 1 \times 2 + 2 \times (-1) + 6 \times 1 \times (-1) \times 2 + 2 \times 1 + (-1)^2 + F_z = 0$$

$$\therefore F_z = 5$$

The body force vector is given by

$$\vec{F} = -11\hat{i} - 16\hat{j} + 5\hat{k}$$

Example 2.12 The rectangular stress components at a point in a three-dimensional stress system are as follows.

$$\sigma_x = 20 \text{ N/mm}^2 \quad \sigma_y = -40 \text{ N/mm}^2 \quad \sigma_z = 80 \text{ N/mm}^2$$
$$\tau_{xy} = 40 \text{ N/mm}^2 \quad \tau_{yz} = -60 \text{ N/mm}^2 \quad \tau_{zx} = 20 \text{ N/mm}^2$$

Determine the principal stresses at the given point.

Solution The principal stresses are the roots of the cubic equation

$$\sigma^3 - I_1\sigma^2 + I_2\sigma - I_3 = 0$$

The three-dimensional stresses can be expressed in the matrix form as below.

$$\begin{bmatrix} \sigma_x & \tau_{xy} & \tau_{xz} \\ \tau_{xy} & \sigma_y & \tau_{yz} \\ \tau_{xz} & \tau_{yz} & \sigma_z \end{bmatrix} = \begin{bmatrix} 20 & 40 & 20 \\ 40 & -40 & -60 \\ 20 & -60 & 80 \end{bmatrix} \text{ N/mm}^2$$

Here,

$$I_1 = (\sigma_x + \sigma_y + \sigma_z)$$
$$= (20 - 40 + 80)$$
$$= 60$$
$$I_2 = \sigma_x\sigma_y + \sigma_y\sigma_z + \sigma_z\sigma_x - \tau_{xy}^2 - \tau_{yz}^2 - \tau_{zx}^2$$
$$= (20(-40) + (-40)(80) + 80(20) - (40)^2 - (-60)^2 - (20)^2)$$
$$= -8000$$
$$I_3 = \sigma_x\sigma_y\sigma_z - \sigma_x\tau_{yz}^2 - \sigma_y\tau_{zx}^2 - \sigma_z\tau_{xy}^2 + 2\tau_{xy}\tau_{yz}\tau_{xz}$$

$$= 20(-40)(80) - (20)(-60)^2 - (-40)(20)^2 - 80(40)^2 + 2(40)(-60)(20)$$
$$= -344{,}000$$

Therefore, cubic equation becomes

$$\sigma^3 - 60\sigma^2 - 8000\sigma + 344{,}000 = 0 \tag{a}$$

Solving the cubic equation for the principal stresses

$$\sigma_1 = 104.98 \text{ N/mm}^2$$
$$\sigma_2 = -83.99 \text{ N/mm}^2$$
$$\sigma_3 = 39.01 \text{ N/mm}^2$$

Example 2.13 At a point in a given material, the three-dimensional state of stress is given by

$$\sigma_x = \sigma_y = \sigma_z = 10 \text{ N/mm}^2, \ \tau_{xy} = 20 \text{ N/mm}^2 \text{ and } \tau_{yz} = \tau_{zx} = 10 \text{ N/mm}^2$$

Compute the principal planes if the corresponding principal stresses are

$$\sigma_1 = 37.3 \text{ N/mm}^2, \quad \sigma_2 = -10 \text{ N/mm}^2, \quad \sigma_3 = 2.7 \text{ N/mm}^2$$

Solution The principal planes can be obtained by their direction Cosines l, m and n associated with each of the three principal stresses, σ_1, σ_2 and σ_3.

(a) To find principal plane for stress σ_1

$$\begin{vmatrix} (10-37.3) & 20 & 10 \\ 20 & (10-37.3) & 10 \\ 10 & 10 & (10-37.3) \end{vmatrix} = \begin{vmatrix} -27.3 & 20 & 10 \\ 20 & -27.3 & 10 \\ 10 & 10 & -27.3 \end{vmatrix}$$

Now,

$$A = \begin{vmatrix} -27.3 & 10 \\ 10 & -27.3 \end{vmatrix} = 745.29 - 100$$

$$A = 645.29$$

$$B = -\begin{vmatrix} 20 & 10 \\ 10 & -27.3 \end{vmatrix}$$

$$= -(-546 - 100)$$

$$B = 646$$

$$C = \begin{vmatrix} 20 & -27.3 \\ 10 & 10 \end{vmatrix}$$

$$= 200 + 270.3$$
$$C = 470.3$$

$$\sqrt{A^2 + B^2 + C^2} = \sqrt{(645.29)^2 + (646)^2 + (470.3)^2}$$
$$= 1027.08$$

$$\therefore l_1 = \frac{A}{\sqrt{A^2 + B^2 + C^2}} = \frac{645.29}{1027.08} = 0.628$$

$$m_1 = \frac{B}{\sqrt{A^2 + B^2 + C^2}} = \frac{646}{1027.08} = 0.628$$

$$n_1 = \frac{C}{\sqrt{A^2 + B^2 + C^2}} = \frac{470.3}{1027.08} = 0.458$$

(b) To find principal plane for stress σ_2

$$\begin{vmatrix} (10+10) & 20 & 10 \\ 20 & (10+10) & 10 \\ 10 & 10 & (10+10) \end{vmatrix} = \begin{vmatrix} 20 & 20 & 10 \\ 20 & 20 & 10 \\ 10 & 10 & 20 \end{vmatrix}$$

$$A = \begin{vmatrix} 20 & 10 \\ 10 & 20 \end{vmatrix} = 400 - 100 = 300$$

$$B = -\begin{vmatrix} 20 & 10 \\ 10 & 20 \end{vmatrix} = -(400 - 100) = -300$$

$$C = \begin{vmatrix} 20 & 20 \\ 10 & 10 \end{vmatrix} = (200 - 200) = 0$$

$$\sqrt{A^2 + B^2 + C^2} = \sqrt{(300)^2 + (-300)^2 + (0)^2} = 424.26$$

$$\therefore l_2 = \frac{A}{\sqrt{A^2 + B^2 + C^2}} = \frac{300}{424.26} = 0.707$$

$$m_2 = \frac{B}{\sqrt{A^2 + B^2 + C^2}} = \frac{-300}{424.26} = -0.707$$

$$n_2 = \frac{C}{\sqrt{A^2 + B^2 + C^2}} = 0$$

(c) To find principal plane for stress σ_3

$$\begin{vmatrix} (10-2.7) & 20 & 10 \\ 20 & (10-2.7) & 10 \\ 10 & 10 & (10-2.7) \end{vmatrix} = \begin{vmatrix} 7.3 & 20 & 10 \\ 20 & 7.3 & 10 \\ 10 & 10 & 7.3 \end{vmatrix}$$

$$A = \begin{vmatrix} 7.3 & 10 \\ 10 & 7.3 \end{vmatrix} = 53.29 - 100 = -46.71$$

$$B = -\begin{vmatrix} 20 & 10 \\ 10 & 7.3 \end{vmatrix} = -(146 - 100) = -46$$

$$C = \begin{vmatrix} 20 & 7.3 \\ 10 & 10 \end{vmatrix} = (200 - 73) = 127$$

$$\sqrt{A^2 + B^2 + C^2} = \sqrt{(-46.71)^2 + (46)^2 + (127)^2} = 142.92$$

$$\therefore l_3 = \frac{A}{\sqrt{A^2 + B^2 + C^2}} = \frac{-46.71}{142.92} = -0.326$$

$$m_3 = \frac{B}{\sqrt{A^2 + B^2 + C^2}} = \frac{-46}{142.92} = -0.322$$

$$n_3 = \frac{C}{\sqrt{A^2 + B^2 + C^2}} = \frac{127}{142.92} = 0.888$$

2.24 Exercises

1. Define stress at a point in a body under the action of external forces.
2. Derive the differential equation of equilibrium in two dimensions.
3. Explain (a) invariants of stress (b) octahedral stresses.
4. What is meant by octahedral shear stress. Arrive at its value in terms of principal stress.
5. Given the following stress matrix (in kN/m^2), obtain the principal stresses and their direction cosines.

$$\begin{bmatrix} 10 & 20 & -40 \\ 20 & -20 & -20 \\ -40 & -20 & 10 \end{bmatrix}$$

6. Explain spherical and deviatoric stress tensor components.
7. If the stress field is given by

$$\sigma_x = \frac{w}{10I}(5x^2 + 2c^2)y - \frac{w}{3I}y^3$$

$$\sigma_y = -\frac{w}{6I}(2c^3 + 3c^2y - y^3)$$

$\tau_{xy} = \frac{w}{2I}x(c^2 - y^2)$, find the body forces required to satisfy the equilibrium conditions.

8. The components of stress at a point are

$$\sigma_x = 2 \text{ MPa} \quad \sigma_y = 1.5 \text{ MPa} \quad \tau_{xy} = \tau_{yz} = 1 \text{ MPa} \quad \tau_{zx} = -1 \text{ MPa}$$

Determine the normal and shearing stresses on the octahedral plane and the direction of the shearing stress.

9. The state of stress at a point is given by the following array of terms in the xyz co-ordinates system

$$\tau_{ij} = \begin{bmatrix} 10 & 15 & 20 \\ 15 & 25 & 15 \\ 20 & 15 & 30 \end{bmatrix} \text{MPa}$$

If this system of axes is rotated by $30°$ about the z-axis in the anticlockwise direction, determine the new stress tensor.

10. The stress components at a point are

$$\sigma_x = -5 \text{ MPa} \quad \sigma_y = 3 \text{ MPa} \quad \sigma_z = 2 \text{ MPa}$$
$$\tau_{xy} = -6 \text{ MPa} \quad \tau_{yz} = 4 \text{ MPa} \quad \tau_{xz} = 5 \text{ MPa}$$

Determine the principal stresses and principal directions.

11. A metal bar is having cross Sect. 2.40 *mm* × 30 *mm*. It is subjected to an axial tensile load of 240 KN. Calculate the normal, shear and resultant stresses on a plane whose normal has the following direction cosines.

(i) $l = m = n = \dfrac{1}{\sqrt{3}}$

(ii) $l = m = \dfrac{1}{\sqrt{2}}$ and $n = 0$

12. The principal stresses at a point in a material are

$$\sigma_1 = 300 \text{ N/mm}^2 \quad \sigma_2 = 300 \text{ N/mm}^2 \quad \text{and} \quad \sigma_3 = 400 \text{ N/mm}^2$$

The direction cosines of the principal planes are $l = m = n = 0.577$. Determine the resultant stress and the invariants.

13. The principal stresses at a point in a three-dimensional stress system are defined as

$$\sigma_1 = 400 \text{ N/mm}^2 \quad \sigma_2 = 200 \text{ N/mm}^2 \quad \text{and} \quad \sigma_3 = 100 \text{ N/mm}^2.$$

Find the normal and shearing stresses on the octahedral planes using Mohr's circle and check the values by theoretical equations.

14. Prove the following relationships

    (i) $(\sigma_n)_{oct} = \dfrac{1}{3}(\sigma_1 + \sigma_2 + \sigma_3)$

    (ii) $\tau_{oct}^2 = \dfrac{1}{9}[(\sigma_1 - \sigma_2)^2 + (\sigma_2 - \sigma_3)^2 + (\sigma_3 - \sigma_1)^2]$

    (iii) $9\tau_{oct}^2 = 2I_1^2 - 6I_2$

15. The state of stress at a point in a body is given by

$$\sigma_x = x^2 y + 20 \quad \tau_{xy} = 3x^2 y$$
$$\sigma_y = x^3 z + y^2 \quad \tau_{yz} = yz$$
$$\sigma_z = yz^2 + 10 \quad \tau_{xz} = xz$$

    Determine the body forces distribution at the point $(1, 2, 3)$ so that the stresses are in equilibrium.

16. The state of stress at a point in a body is given with reference axes as

$$\sigma_x = 200 \text{ N/m}^2 \quad \tau_{xy} = 100 \text{ N/m}^2$$
$$\sigma_y = 0 \qquad\qquad \tau_{yz} = 0$$
$$\sigma_z = 500 \text{ N/m}^2 \quad \tau_{xz} = 0$$

    if a new set of axis $x'\, y'\, z'$ is formed by rotating $x\, y\, z$ axes through $60°$ about the z-axis in the anticlockwise direction, determine the components of stress for the new axes. Also prove that the invariants remain unchanged.

17. The components of stress at a point are:

$$\sigma_x = 10 \text{ kPa} \quad \tau_{xy} = 20 \text{ kPa}$$
$$\sigma_y = -20 \text{ kPa} \quad \tau_{yz} = 30 \text{ kPa}$$
$$\sigma_z = -20 \text{ kPa} \quad \tau_{xz} = 30 \text{ kPa}$$

    Determine

    (a) The principal stresses at the point
    (b) Deviatoric and spherical stress tensors.

18. The stress components at a point in cylindrical co-ordinates are:

$$\sigma_r = r^3\theta + r \qquad \tau_{r\theta} = r^2\theta$$
$$\sigma_\theta = r^2 z + \theta^2 \quad \tau_{\theta z} = \theta z + \theta^2$$
$$\sigma_z = r^2 z^2 + \theta z \quad \tau_{rz} = rz^2$$

Determine the body force distributions at the point $(3, \frac{\pi}{3}, 5)$ for the body to be in equilibrium.

19. The state of stress at a point in a body is specified by the following stress components:

$$\begin{aligned}
\sigma_x &= 110 \text{ MPa} \quad \tau_{xy} = 60 \text{ MPa} \\
\sigma_y &= -86 \text{ MPa} \quad \tau_{yz} = \tau_{zx} = 0 \\
\sigma_z &= 55 \text{ MPa}
\end{aligned}$$

Determine the principal stresses, direction cosines of the principal stress directions and the maximum shearing stress.

Chapter 3
Analysis of Strain

3.1 Deformations

In Chap. 2, the effect of externally applied forces on the body was discussed in terms of stresses developed within the material concerned. Usually, these stresses may also cause the body to transform into a new configuration. Hence, the change in the shape of the body from initial and final configurations is termed as deformation.

3.2 Displacement and Strain

Deformations occur mainly due to relative displacements between the points in a body. Suppose, P and Q are two points in a body at a distance L_o apart before deformation, as shown in Fig. 3.1. Now let this body be acted upon by a combination of external forces causing the body to take up the new position in which the points P and Q move to P' and Q'. Here, the distance PP' through which P has moved is called the displacement of P, and Q Q' is the displacement of Q. If P' Q' is parallel to PQ, then the deformation in this case has been translation only. But, if points P and Q have moved to P' and Q' in such a way that $P'Q''$ is not parallel to PQ, then the deformation includes both translation and rotation.

Further, if the distance L_1 between P' and Q' is equal to L_o then the line PQ has been rigidly moved and no straining has taken place in the line element PQ and thus also does not affect the stresses. However, if the distance L_1 is not equal to L_o, then there has been displacement of Q relative to P and a state of strain has been set up.

The position and displacement of any point in the body can be specified with respect to any co-ordinate system such as the Cartesian set x, y, z or the cylindrical set r, θ and z. Thus, in the two-dimensional case as shown in Fig. 3.1, the point P has co-ordinates (x_P, y_P). The components of the displacement of P to P' can be represented by u_P and v_P along the co-ordinate directions x and y, respectively.

© The Author(s), under exclusive license to Springer Nature Singapore Pte Ltd. 2021   55
T. G. Sitharam and L. Govindaraju, *Theory of Elasticity*,
https://doi.org/10.1007/978-981-33-4650-5_3

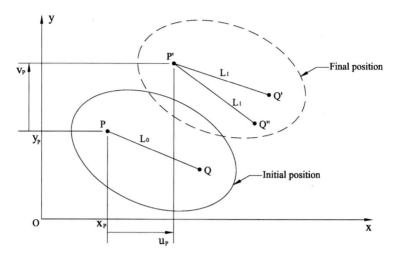

Fig. 3.1 Displacement and strain of a body

Generally, the symbols u, v and w are used to represent displacement components in x, y and z directions, respectively.

3.3   Concept of Strain

As stated earlier, deformations occur due to relative displacements between points in a body. However, if the shape and size of the body remain unaltered, then this type of deformation is known as rigid body motion. But if the shape and size of the body are altered, then a state of strain has been set up in the body. Generally, the strain will not be constant or homogeneous with in a body but will be different at different points due to the variation in the displacements from particle to particle. But if the region of the material of the body considered is sufficiently small, then the strain within it may be treated as constant or homogeneous. Thus, the strain along $P'\,Q'$ in Fig. 3.1 may be defined as

$$\varepsilon = \frac{L_1 - L_o}{L_o} = \frac{L_1}{L_o} - 1 \tag{3.1}$$

The change in length $(L_1 - L_o)$ with the original length L_o is known as strain.

3.4 Components of Strain

In a manner similar to the state of stress at any point in a body, as discussed in Chap. 2, in order to specify the state of strain at any point, it becomes necessary to define it in terms of relative movement of adjacent points located along the three orthogonal axes at a point. For the sake of clarity, it is necessary to describe the relative movement of adjacent points in terms of infinitesimal line segments connecting these points to be the sides of an infinitesimal rectangular parallelepiped. Let this rectangular parallelepiped has the sides initially normal to the three mutually perpendicular axes X, Y and Z as shown in Fig. 3.2.

Now, let us consider each side of the parallelepiped before and after deformation with reference to the three mutually perpendicular planes XY, YZ and ZX.

Referring to Fig. 3.3 let the side ABCD is projected on to the plane XOY. This side after deformation is represented by $A'\ B'\ C'\ D'$. From Fig. 3.3, it is observed that the displacement of A after deformation is "u" and the displacement of C is $(u + \partial u)$ considering the variation in lengths between $A'\ C'$ and AC caused by the deformation. But the intensity of variation in length is $(\partial u/\partial x)$ and the total increment in length along X-axis is $(\partial u/\partial x)\ \Delta x$. Hence, $(\partial u/\partial x)\ \Delta x$ is the component of relative displacement of C with respect to A or the projection of side $A'\ C'$ on the X-axis $(A'\ C'')$. Similarly, $(\partial v/\partial x)\ \Delta x$ is the angular displacement of C with respect to A or the projection of side $C'\ C''$ on the Y-axis.

Now, the relative elongation of side AC parallel to X-axis is:

Fig. 3.2 Rectangular parallelepiped in space co-ordinates

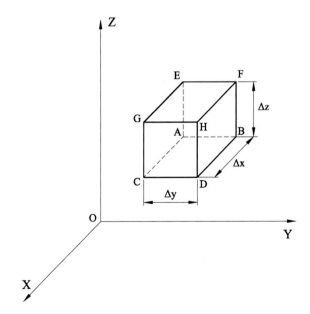

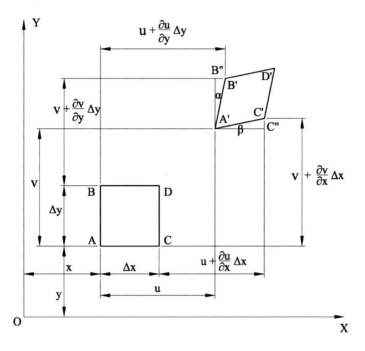

Fig. 3.3 Two-dimensional geometric strain deformation

$$\varepsilon_x = \frac{A'C'' - AC}{AC} = \frac{[u + (\partial u/\partial x)\,\Delta x + \Delta x - u] - \Delta x}{\Delta x} = \partial u/\partial x \qquad (3.2)$$

Similarly, if v and w are the displacement components in Y and Z directions, respectively, then the strain components in those directions derived as

$$\varepsilon_y = \partial v/\partial y \quad \varepsilon_z = \partial w/\partial y \qquad (3.3)$$

In order to determine the angular deformation, let us consider the rotation of side AC with respect to A from x-axis.

Hence,

$$\tan \beta = \frac{C'C''}{A'C''}$$
$$\tan \beta = \frac{(v + \partial v/\partial x\,\Delta x) - v}{(u + \partial u/\partial x\,\Delta x) + \Delta x - u}$$
$$\tan \beta = \frac{(\partial v/\partial x)}{1 + (\partial u/\partial x)}$$

Considering only infinitesimal deformations, then $\partial u/\partial x$ is very small compared to unity.

Hence,

$$\tan \beta = \partial v / \partial x \approx \beta \tag{3.4}$$

Similarly,

$$\tan \alpha = \partial u / \partial y \approx \alpha \tag{3.5}$$

Therefore, the change in angle between the two sides AB and AC due to deformation is:

$$\gamma_{xy} = \alpha + \beta = (\partial u / \partial y + \partial v / \partial x) \tag{3.6}$$

Similarly, in other two directions, it can be proved that

$$\gamma_{yz} = (\partial v / \partial z + \partial w / \partial y) \tag{3.7}$$

$$\gamma_{zx} = (\partial w / \partial x + \partial u / \partial z) \tag{3.8}$$

Here, ε_x, ε_y and ε_z denote normal strains in X, Y and Z directions and γ_{xy}, γ_{yz} and γ_{zx} denote shear strains in XY, YZ and ZX planes.

3.5 Strain Tensor

Just as the state of stress at a point is described by nine-term array, the strain can be represented tensorially as below:

$$\varepsilon_{ij} = \frac{1}{2} \left[\frac{\partial u_i}{\partial x_j} + \frac{\partial u_j}{\partial x_i} \right] \quad (i, \ j = x, \ y, \ z) \tag{3.9}$$

The factor 1/2 in Eq. (3.9) is used to represent the strain transformation equations in indicial notation. The longitudinal strains are obtained when $i = j$ and the shear strains are obtained when $i \neq j$.

Therefore, it can be expressed as

$$\varepsilon_{xy} = \frac{1}{2} \gamma_{xy}, \ \varepsilon_{yz} = \frac{1}{2} \gamma_{yz} \ \text{and} \ \varepsilon_{zx} = \frac{1}{2} \gamma_{zx} \tag{3.10}$$

Hence, the corresponding strain tensor is given by

$$\varepsilon_{ij} = \begin{bmatrix} \varepsilon_x & \varepsilon_{xy} & \varepsilon_{xz} \\ \varepsilon_{yx} & \varepsilon_y & \varepsilon_{yz} \\ \varepsilon_{zx} & \varepsilon_{zy} & \varepsilon_z \end{bmatrix} \tag{3.11}$$

$$\varepsilon_{ij} = \begin{bmatrix} \varepsilon_x & \frac{1}{2}\gamma_{xy} & \frac{1}{2}\gamma_{xz} \\ \frac{1}{2}\gamma_{yx} & \varepsilon_y & \frac{1}{2}\gamma_{yz} \\ \frac{1}{2}\gamma_{zx} & \frac{1}{2}\gamma_{zy} & \varepsilon_z \end{bmatrix} \tag{3.12}$$

3.6 Rotations

From Fig. 3.2, it is clear that, for infinitesimally small strains, the angle of rotation of side $A'\ C'$ from X-axis towards Y-axis is $(\partial v/\partial x)$ in anticlockwise direction. Similarly, the angle of rotation the side $A'\ B'$ from Y-axis towards X-axis is $(\partial u/\partial y)$ in clockwise direction. Therefore, the sum of these two displacement gradients gives the total relative rotation of two sides. Hence, the shear strains as in (3.6), the difference between them gives the total rotation of the element in the X, Y plane as below.

$$\omega_{xy} = \left(\frac{\partial u}{\partial y} - \frac{\partial v}{\partial x} \right) \tag{3.13}$$

This is the rigid body rotation of the element about an axis parallel to the Z-axis. In similar manner to the definition of shear strains, the definition for rotation would take the average rotation of the two sides. Hence, rotation of element about an axis parallel to Z-axis is given by:

$$\omega_{xy} = \frac{1}{2}\left(\frac{\partial u}{\partial y} - \frac{\partial v}{\partial x} \right) \tag{3.14}$$

The rotations about the X and Y axes may be similarly defined as

$$\omega_{yz} = \frac{1}{2}\left(\frac{\partial v}{\partial z} - \frac{\partial w}{\partial y} \right) \tag{3.15}$$

$$\omega_{zx} = \frac{1}{2}\left(\frac{\partial w}{\partial x} - \frac{\partial u}{\partial z} \right) \tag{3.16}$$

3.7 Deformation of an Infinitesimal Line Element

Consider an infinitesimal line element PQ in the undeformed geometry of a medium as shown in Fig. 3.4. When the body undergoes deformation, the line element PQ passes into the line element $P'Q'$. In general, both the length and the direction of PQ are changed.

Let the co-ordinates of P and Q before deformation be (x, y, z), $(x + \Delta x, y + \Delta y, z + \Delta z)$, respectively, and the displacement vector at point P has components (u, v, w). The co-ordinates of P, P' and Q are

$$P : (x, y, z)$$
$$P' : (x + u, y + v, z + w)$$
$$Q : (x + \Delta x, y + \Delta y, z + \Delta z)$$

The displacement components at Q differ slightly from those at point P since Q is away from P by Δx, Δy and Δz. Hence the displacements at Q are

$$u + \Delta u, v + \Delta v \quad \text{and} \quad w + \Delta w$$

Now, if Q is very close to P, then to the first-order approximation

$$\Delta u = \frac{\partial u}{\partial x}\Delta x + \frac{\partial u}{\partial y}\Delta y + \frac{\partial u}{\partial z}\Delta z \tag{3.17}$$

Fig. 3.4 Line element in undeformed and deformed body

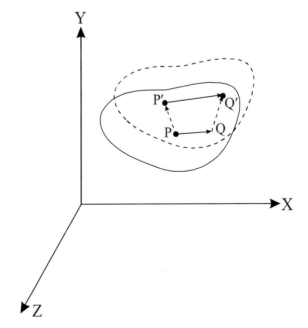

Similarly,

$$\Delta v = \frac{\partial v}{\partial x}\Delta x + \frac{\partial v}{\partial y}\Delta y + \frac{\partial v}{\partial z}\Delta z \tag{3.18}$$

And

$$\Delta w = \frac{\partial w}{\partial x}\Delta x + \frac{\partial w}{\partial y}\Delta y + \frac{\partial w}{\partial z}\Delta z \tag{3.19}$$

The co-ordinates of Q' are, therefore,

$$Q'(x + \Delta x + u + \Delta u, \; y + \Delta y + v + \Delta v, \; z + \Delta z + w + \Delta w)$$

Before deformation, the segment PQ had components Δx, Δy and Δz along the three axes.

After deformation, the segment $P'Q'$ has components $\Delta x + u$, $\Delta y + v$ and $\Delta z + w$ along the three axes.

Here, the terms like $\frac{\partial u}{\partial x}$, $\frac{\partial u}{\partial y}$ and $\frac{\partial u}{\partial z}$, etc., are important in the analysis of strain. These are the gradients of the displacement components in x, y and z directions. These can be represented in the form of a matrix called the displacement-gradient matrix such as

$$\left[\frac{\partial u_i}{\partial x_j} \right] = \begin{bmatrix} \frac{\partial u}{\partial x} & \frac{\partial u}{\partial y} & \frac{\partial u}{\partial z} \\ \frac{\partial v}{\partial x} & \frac{\partial v}{\partial y} & \frac{\partial v}{\partial z} \\ \frac{\partial w}{\partial x} & \frac{\partial w}{\partial y} & \frac{\partial w}{\partial z} \end{bmatrix}$$

3.8   Change in Length of a Linear Element

When the body undergoes deformation, it causes a point $P(x, \; y, \; z)$ in the body under consideration to be displaced to a new position P' with co-ordinates $(x + u, \; y + v, \; z + w)$ where u, v and w are the displacement components. Also, a neighbouring point Q with co-ordinates $(x + \Delta x, \; y + \Delta y, \; z + \Delta z)$ gets displaced to Q' with new co-ordinates $(x + \Delta x + u + \Delta u, \; y + \Delta y + v + \Delta v, \; z + \Delta z + w + \Delta w)$.

Now, let ΔS be the length of the line element PQ with its components $(\Delta x, \; \Delta y, \; \Delta z)$.

Hence,

$$(\Delta S)^2 = (PQ)^2 = (\Delta x)^2 + (\Delta y)^2 + (\Delta z)^2$$

Similarly, $\Delta S'$ be the length $P'Q'$ with its components

$$\left(\Delta x' = \Delta x + \Delta u, \ \Delta y' = \Delta y + \Delta v, \ \Delta z' = \Delta z + \Delta w\right)$$

$$\therefore \left(\Delta S'\right)^2 = \left(P'Q'\right)^2 = (\Delta x + \Delta u)^2 + (\Delta y + \Delta v)^2 + (\Delta z + \Delta w)^2$$

From Eqs. (3.17)–(3.19),

$$\Delta x' = \left(1 + \frac{\partial u}{\partial x}\right)\Delta x + \frac{\partial u}{\partial y}\Delta y + \frac{\partial u}{\partial z}\Delta z$$
$$\Delta y' = \frac{\partial v}{\partial x}\Delta x + \left(1 + \frac{\partial v}{\partial y}\right)\Delta y + \frac{\partial v}{\partial z}\Delta z$$
$$\Delta z' = \frac{\partial w}{\partial x}\Delta x + \frac{\partial w}{\partial y}\Delta y + \left(1 + \frac{\partial w}{\partial z}\right)\Delta z$$

Taking the difference between $\left(\Delta S'\right)^2$ and $(\Delta S)^2$, we get

$$\begin{aligned}
\left(P'Q'\right)^2 - (PQ)^2 &= \left(\Delta S'\right)^2 - (\Delta S)^2 \\
&\left\{ \left((\Delta x')^2 + \left(\Delta y'\right)^2 + \left(\Delta z'\right)^2\right) - \left((\Delta x)^2 + (\Delta y)^2 + (\Delta z)^2\right) \right\} \\
&= 2\left(\varepsilon_x \Delta x^2 + \varepsilon_y \Delta y^2 + \varepsilon_z \Delta z^2 + \varepsilon_{xy}\Delta x \Delta y + \varepsilon_{yz}\Delta y \Delta z + \varepsilon_{zx}\Delta x \Delta z\right)
\end{aligned} \tag{3.20}$$

where

$$\varepsilon_x = \frac{\partial u}{\partial x} + \frac{1}{2}\left[\left(\frac{\partial u}{\partial x}\right)^2 + \left(\frac{\partial v}{\partial x}\right)^2 + \left(\frac{\partial w}{\partial x}\right)^2\right] \tag{3.20a}$$

$$\varepsilon_y = \frac{\partial v}{\partial y} + \frac{1}{2}\left[\left(\frac{\partial u}{\partial y}\right)^2 + \left(\frac{\partial v}{\partial y}\right)^2 + \left(\frac{\partial w}{\partial y}\right)^2\right] \tag{3.20b}$$

$$\varepsilon_z = \frac{\partial w}{\partial z} + \frac{1}{2}\left[\left(\frac{\partial u}{\partial z}\right)^2 + \left(\frac{\partial v}{\partial z}\right)^2 + \left(\frac{\partial w}{\partial z}\right)^2\right] \tag{3.20c}$$

$$\varepsilon_{xy} = \varepsilon_{yx} = \left[\frac{\partial v}{\partial x} + \frac{\partial u}{\partial y} + \frac{\partial u}{\partial x}\frac{\partial u}{\partial y} + \frac{\partial v}{\partial x}\frac{\partial v}{\partial y} + \frac{\partial w}{\partial x}\frac{\partial w}{\partial y}\right] \tag{3.20d}$$

$$\varepsilon_{yz} = \varepsilon_{zy} = \left[\frac{\partial w}{\partial y} + \frac{\partial v}{\partial z} + \frac{\partial u}{\partial y}\frac{\partial u}{\partial z} + \frac{\partial v}{\partial y}\frac{\partial v}{\partial z} + \frac{\partial w}{\partial y}\frac{\partial w}{\partial z}\right] \tag{3.20e}$$

$$\varepsilon_{zx} = \varepsilon_{xz} = \left[\frac{\partial u}{\partial z} + \frac{\partial w}{\partial x} + \frac{\partial u}{\partial z}\frac{\partial u}{\partial x} + \frac{\partial v}{\partial z}\frac{\partial v}{\partial x} + \frac{\partial w}{\partial z}\frac{\partial w}{\partial x}\right] \tag{3.20f}$$

Now, introducing the notation

$$\varepsilon_{PQ} = \frac{\Delta S' - \Delta S}{\Delta S}$$

which is called the relative extension of point P in the direction of point Q, now,

$$\frac{(\Delta S')^2 - (\Delta S)^2}{2} = \left(\frac{\Delta S' - \Delta S}{\Delta S} + \frac{(\Delta S' - \Delta S)^2}{2(\Delta S)^2}\right)(\Delta S)^2$$

$$= \left[\varepsilon_{PQ} + \frac{1}{2}(\varepsilon_{PQ})^2\right](\Delta S)^2$$

$$= \varepsilon_{PQ}\left[1 + \frac{1}{2}\varepsilon_{PQ}\right](\Delta S)^2$$

From Eq. (3.20), substituting for $(\Delta S')^2 - (\Delta S)^2$, we get

$$\varepsilon_{PQ}\left(1 + \frac{1}{2}\varepsilon_{PQ}\right)(\Delta S)^2 = \varepsilon_x(\Delta x)^2 + \varepsilon_y(\Delta y)^2 + \varepsilon_z(\Delta z)^2 + \varepsilon_{xy}\Delta x\Delta y$$

$$+ \varepsilon_{yz}\Delta y\Delta z + \varepsilon_{zx}\Delta x\Delta z$$

If l, m and n are the direction cosines of PQ, then

$$l = \frac{\Delta x}{\Delta S}, \quad m = \frac{\Delta y}{\Delta S}, \quad n = \frac{\Delta z}{\Delta S}$$

Substituting these quantities in the above expression,

$$\varepsilon_{PQ}\left(1 + \frac{1}{2}\varepsilon_{PQ}\right) = \varepsilon_x l^2 + \varepsilon_y m^2 + \varepsilon_z n^2 + \varepsilon_{xy}lm + \varepsilon_{yz}mn + \varepsilon_{zx}nl$$

The above equation gives the value of the relative displacement at point P in the direction PQ with direction cosines l, m and n.

3.9   Change in Length of a Linear Element—Linear Components

It can be observed from Eqs. (3.20a)–(3.20c) that they contain linear terms like $\frac{\partial u}{\partial x}, \frac{\partial v}{\partial y}, \frac{\partial w}{\partial z}, ----$etc., as well as nonlinear terms like $\left(\frac{\partial u}{\partial x}\right)^2$, $\left(\frac{\partial u}{\partial x}\cdot\frac{\partial u}{\partial y}\right)$, $----$etc. If the deformation imposed on the body is small, the terms like $\frac{\partial u}{\partial x}$, $\frac{\partial v}{\partial y}$, etc are extremely small so that their squares and products can be neglected. Hence, retaining only linear terms, the linear strain at point P in the direction PQ can be obtained as below.

$$\varepsilon_x = \frac{\partial u}{\partial x}, \quad \varepsilon_y = \frac{\partial v}{\partial y}, \quad \varepsilon_z = \frac{\partial w}{\partial z} \tag{3.21a}$$

$$\gamma_{xy} = \frac{\partial u}{\partial y} + \frac{\partial v}{\partial x}, \quad \gamma_{yz} = \frac{\partial v}{\partial z} + \frac{\partial w}{\partial y}, \quad \gamma_{zx} = \frac{\partial w}{\partial x} + \frac{\partial u}{\partial z} \tag{3.21b}$$

and $\varepsilon_{PQ} \cong \varepsilon_{PQ} = \varepsilon_x l^2 + \varepsilon_y m^2 + \varepsilon_z n^2 + \gamma_{xy}\, lm + \gamma_{yz}\, mn + \gamma_{zx}\, nl$ \qquad (3.22)

If, however, the line element is parallel to X-axis, then $l = 1, m = 0, n = 0$ and the linear strain is

$$\varepsilon_{PQ} = \varepsilon_x = \frac{\partial u}{\partial x}$$

Similarly, for element parallel to Y-axis, then $l = 0, m = 1, n = 0$ and the linear strain is

$$\varepsilon_{PQ} = \varepsilon_y = \frac{\partial v}{\partial y}$$

and for element parallel to z-axis, then $l = 0, m = 0, n = 1$ and the linear strain is

$$\varepsilon_{PQ} = \varepsilon_z = \frac{\partial w}{\partial z}$$

The relations expressed by Eqs. (3.21a) and (3.21b) are known as the strain–displacement relations of Cauchy.

3.10   Strain Transformation

If the displacement components u, v and w at a point are represented in terms of known functions of x, y and z, respectively, in Cartesian co-ordinates, then the six strain components can be determined by using the strain–displacement relations given below.

$$\varepsilon_x = \frac{\partial u}{\partial x}, \qquad \varepsilon_y = \frac{\partial v}{\partial y}, \qquad \varepsilon_z = \frac{\partial w}{\partial z}$$

$$\gamma_{xy} = \frac{\partial u}{\partial y} + \frac{\partial v}{\partial x}, \qquad \gamma_{yz} = \frac{\partial v}{\partial z} + \frac{\partial w}{\partial y} \quad \text{and} \quad \gamma_{zx} = \frac{\partial w}{\partial x} + \frac{\partial u}{\partial z}$$

If at the same point, the strain components with reference to another set of co-ordinates axes x', y' and z' are desired, then they can be calculated using the concepts of axis transformation and the corresponding direction cosines. It is to be noted that the above equations are valid for any system of orthogonal co-ordinate axes irrespective of their orientations.

Hence,

$$\varepsilon_{x'} = \frac{\partial u'}{\partial x'}, \qquad \varepsilon_{y'} = \frac{\partial v'}{\partial y'}, \qquad \varepsilon_{z'} = \frac{\partial w'}{\partial z'}$$

$$\gamma_{x'y'} = \frac{\partial u'}{\partial y'} + \frac{\partial v'}{\partial x'}, \quad \gamma_{y'z'} = \frac{\partial v'}{\partial z'} + \frac{\partial w'}{\partial y'}, \quad \gamma_{z'x'} = \frac{\partial w'}{\partial x'} + \frac{\partial u'}{\partial z'}$$

Similar to the transformation of stresses, the transformation of strains from one co-ordinate system to another can be written in matrix for as below.

$$\begin{bmatrix} \varepsilon_{x'} & \frac{1}{2}\gamma_{x'y'} & \frac{1}{2}\gamma_{x'z'} \\ \frac{1}{2}\gamma_{y'x'} & \varepsilon_{y'} & \frac{1}{2}\gamma_{y'z'} \\ \frac{1}{2}\gamma_{z'x'} & \frac{1}{2}\gamma_{z'y'} & \varepsilon_{z'} \end{bmatrix} = \begin{bmatrix} l_1 & m_1 & n_1 \\ l_2 & m_2 & n_2 \\ l_3 & m_3 & n_3 \end{bmatrix} \begin{bmatrix} \varepsilon_x & \frac{1}{2}\gamma_{xy} & \frac{1}{2}\gamma_{xz} \\ \frac{1}{2}\gamma_{yx} & \varepsilon_y & \frac{1}{2}\gamma_{yz} \\ \frac{1}{2}\gamma_{zx} & \frac{1}{2}\gamma_{zy} & \varepsilon_z \end{bmatrix} \begin{bmatrix} l_1 & l_2 & l_3 \\ m_1 & m_2 & m_3 \\ n_1 & n_2 & n_3 \end{bmatrix}$$

In general, $\left[\varepsilon'\right] = [a] \, [\varepsilon] \, [a]^T$.

3.11   Spherical and Deviatorial Strain Tensors

Like the stress tensor, the strain tensor is also divided into two parts, the spherical and the deviatorial as:

$$E = E'' + E'$$

$$\text{where } E'' = \begin{bmatrix} e & 0 & 0 \\ 0 & e & 0 \\ 0 & 0 & e \end{bmatrix} = \text{spherical strain} \qquad (3.23)$$

$$E' = \begin{bmatrix} (\varepsilon_x - e) & \varepsilon_{xy} & \varepsilon_{xz} \\ \varepsilon_{yx} & (\varepsilon_y - e) & \varepsilon_{yz} \\ \varepsilon_{zx} & \varepsilon_{xy} & (\varepsilon_z - e) \end{bmatrix} = \text{deviatorial strain} \qquad (3.24)$$

and $e = \frac{\varepsilon_x + \varepsilon_y + \varepsilon_z}{3}$.

It is noted that the spherical component E'' produces only volume changes without any change of shape while the deviatorial component E' produces distortion or change of shape. These components are extensively used in theories of failure and are sometimes known as "dilatation" and "distortion" components.

3.12 Principal Strains and Strain Invariants

During the discussion of the state of stress at a point, it was stated that at any point in a continuum there exists three mutually orthogonal planes, known as principal planes, on which there are no shear stresses.

Similar to that, planes exist on which there are no shear strains and only normal strains occur. These planes are termed as principal planes and the corresponding strains known as principal strains. The principal strains can be obtained by first determining the three mutually perpendicular directions along which the normal strains have stationary values. Hence, for this purpose, the normal strains given by Eq. (3.22) can be used.

$$\text{i.e.,}\quad \varepsilon_{PQ} = \varepsilon_x l^2 + \varepsilon_y m^2 + \varepsilon_z n^2 + \gamma_{xy} lm + \gamma_{yz} mn + \gamma_{zx} nl$$

As the values of l, m and n change, one can get different values for the strain ε_{PQ}. Therefore, to find the maximum or minimum values of strain, we are required to equate $\frac{\partial \varepsilon_{PQ}}{\partial l}$, $\frac{\partial \varepsilon_{PQ}}{\partial m}$, $\frac{\partial \varepsilon_{PQ}}{\partial n}$ to zero, if l, m and n were all independent. But, l, m and n are not at all independent, since they are related by the relation.

$$l^2 + m^2 + n^2 = 1$$

Now, taking l and m as independent and differentiating with respect to l and m, we get

$$2l + 2n \frac{\partial n}{\partial l} = 0$$
$$2m + 2n \frac{\partial n}{\partial m} = 0 \tag{3.25}$$

Now differentiating ε_{PQ} with respect to l and m for an extremum, we get

$$0 = 2l\varepsilon_x + m\gamma_{xy} + n\gamma_{zx} + \frac{\partial n}{\partial l}\left(l\gamma_{zx} + m\gamma_{zy} + 2n\varepsilon_z\right)$$

$$0 = 2m\varepsilon_y + l\gamma_{xy} + n\gamma_{yz} + \frac{\partial n}{\partial m}\left(l\gamma_{zx} + m\gamma_{zy} + 2n\varepsilon_z\right)$$

Substituting for $\frac{\partial n}{\partial l}$ and $\frac{\partial n}{\partial m}$ from Eq. (iv), we get

$$\frac{2l\varepsilon_x + m\gamma_{xy} + n\gamma_{zx}}{l} = \frac{l\gamma_{zx} + m\gamma_{zy} + 2n\varepsilon_z}{n}$$
$$\frac{2m\varepsilon_y + l\gamma_{xy} + n\gamma_{yz}}{m} = \frac{l\gamma_{zx} + m\gamma_{zy} + 2n\varepsilon_z}{n}$$

Denoting the right-hand expression in the above two equations by 2ε,

$$2\varepsilon_x l + \gamma_{xy} m + \gamma_{xz} n - 2\varepsilon l = 0$$
$$\gamma_{xy} l + 2\varepsilon_y m + \gamma_{yz} n - 2\varepsilon m = 0 \tag{3.25a}$$
$$\text{and } \gamma_{zx} l + \gamma_{zy} m + 2\varepsilon_z n - 2\varepsilon n = 0$$

Using Eq. (3.25a), we can obtain the values of l, m and n which determine the direction along which the relative extension is an extremum. Now, multiplying the first Equation by l, the second by m and the third by n, and adding them,
We get

$$2\left(\varepsilon_x l^2 + \varepsilon_y m^2 + \varepsilon_z n^2 + \gamma_{xy} lm + \gamma_{yz} mn + \gamma_{zx} nl\right) = 2\varepsilon\left(l^2 + m^2 + n^2\right) \quad (3.25b)$$

Here, $\varepsilon_{PQ} = \varepsilon_x l^2 + \varepsilon_y m^2 + \varepsilon_z n^2 + \gamma_{xy} lm + \gamma_{yz} mn + \gamma_{zx} nl$

$$l^2 + m^2 + n^2 = 1$$

Hence, Eq. (3.25b) can be written as

$$\varepsilon_{PQ} = \varepsilon$$

which means that in Eq. (3.25a), the values of l, m and n determine the direction along which the relative extension is an extremum and also, the value of ε is equal to this extremum. Hence, Eq. (3.25a) can be written as

$$\begin{aligned}
(\varepsilon_x - \varepsilon) l + \tfrac{1}{2}\gamma_{xy} m + \tfrac{1}{2}\gamma_{xz} n &= 0 \\
\tfrac{1}{2}\gamma_{yx} l + \left(\varepsilon_y - \varepsilon\right) m + \tfrac{1}{2}\gamma_{yz} n &= 0 \\
\tfrac{1}{2}\gamma_{zx} l + \tfrac{1}{2}\gamma_{zy} m + (\varepsilon_z - \varepsilon) n &= 0
\end{aligned} \quad (3.25c)$$

Denoting
$\tfrac{1}{2}\gamma_{xy} = \varepsilon_{xy}$, $\tfrac{1}{2}\gamma_{yz} = \varepsilon_{yz}$, $\tfrac{1}{2}\gamma_{zx} = \varepsilon_{zx}$, then
Equation (3.25c) can be written as

$$\begin{aligned}
(\varepsilon_x - \varepsilon) + \varepsilon_{xy} m + \varepsilon_{xz} n &= 0 \\
\varepsilon_{yx} l + \left(\varepsilon_y - \varepsilon\right) m + \varepsilon_{yz} n &= 0 \\
\varepsilon_{zx} l + \varepsilon_{zy} m + (\varepsilon_z - \varepsilon) n &= 0
\end{aligned} \quad (3.25d)$$

The above set of equations is homogenous in l, m and n. In order to obtain a non-trivial solution of the directions l, m and n from Eq. (3.25d), the determinant of the coefficients should be zero.

i.e.,
$$\begin{vmatrix}
(\varepsilon_x - \varepsilon) & \varepsilon_{xy} & \varepsilon_{xz} \\
\varepsilon_{yx} & (\varepsilon_y - \varepsilon) & \varepsilon_{yz} \\
\varepsilon_{zx} & \varepsilon_{zy} & (\varepsilon_z - \varepsilon)
\end{vmatrix} = 0$$

Expanding the determinant of the coefficients, we get

$$\varepsilon^3 - J_1 \varepsilon^2 + J_2 \varepsilon - J_3 = 0 \quad (3.25e)$$

where

$$J_1 = \varepsilon_x + \varepsilon_y + \varepsilon_z$$

$$J_2 = \begin{vmatrix} \varepsilon_x & \varepsilon_{xy} \\ \varepsilon_{yx} & \varepsilon_y \end{vmatrix} + \begin{vmatrix} \varepsilon_y & \varepsilon_{yz} \\ \varepsilon_{zy} & \varepsilon_z \end{vmatrix} + \begin{vmatrix} \varepsilon_z & \varepsilon_{zx} \\ \varepsilon_{xz} & \varepsilon_x \end{vmatrix}$$

$$J_3 = \begin{vmatrix} \varepsilon_x & \varepsilon_{xy} & \varepsilon_{xz} \\ \varepsilon_{yx} & \varepsilon_y & \varepsilon_{yz} \\ \varepsilon_{zx} & \varepsilon_{zy} & \varepsilon_z \end{vmatrix}$$

We can also write as

$$J_1 = \varepsilon_x + \varepsilon_y + \varepsilon_z$$

$$J_2 = \varepsilon_x \varepsilon_y + \varepsilon_y \varepsilon_z + \varepsilon_z \varepsilon_x - \frac{1}{4}\left(\gamma_{xy}^2 + \gamma_{yz}^2 + \gamma_{zx}^2\right)$$

$$J_3 = \varepsilon_x \varepsilon_y \varepsilon_z + \frac{1}{4}\left(\gamma_{xy}\gamma_{yz}\gamma_{zx} - \varepsilon_x \gamma_{yz}^2 - \varepsilon_y \gamma_{zx}^2 - \varepsilon_z \gamma_{xy}^2\right)$$

Hence, the three roots ε_1, ε_2 and ε_3 of the cubic Eq. (3.25e) are known as the principal strains and J_1, J_2 and J_3 are termed as first invariant, second invariant and third invariant of strains, respectively.

3.13 Octahedral Strains

The strains acting on a plane which is equally inclined to the three co-ordinate axes are known as octahedral strains. The direction cosines of the normal to the octahedral plane are, $\frac{1}{\sqrt{3}}, \frac{1}{\sqrt{3}}, \frac{1}{\sqrt{3}}$.

The normal octahedral strain is:

$$(\varepsilon_n)_{\text{oct}} = \varepsilon_1 \, l^2 + \varepsilon_2 m^2 + \varepsilon_3 n^2$$

$$\therefore (\varepsilon_n)_{\text{oct}} = \frac{1}{3}(\varepsilon_1 + \varepsilon_2 + \varepsilon_3) \tag{3.26}$$

$$\text{Resultant octahedral strain} = (e_R)_{\text{oct}} = \sqrt{(\varepsilon_1 l)^2 + (\varepsilon_2 m)^2 + (\varepsilon_3 n)^2}$$

$$= \sqrt{\frac{1}{3}\left(\varepsilon_1^2 + \varepsilon_2^2 + \varepsilon_3^2\right)} \tag{3.27}$$

$$\text{Octahedral shear strain} = \gamma_{\text{oct}} = \frac{2}{3}\sqrt{(\varepsilon_1 - \varepsilon_2)^2 + (\varepsilon_2 - \varepsilon_3)^2 + (\varepsilon_3 - \varepsilon_1)^2} \tag{3.28}$$

3.14 Mohr's Circle for Strain

The Mohr's circle for strain is drawn and that the construction technique does not differ from that of Mohr's circle for stress. In Mohr's circle for strain, the normal strains are plotted on the horizontal axis, positive to right. When the shear strain is positive, the point representing the x-axis strains is plotted at a distance $\frac{\gamma}{2}$ below the ε-line; and the y-axis point a distance $\frac{\gamma}{2}$ above the ε-line; and vice versa when the shear strain is negative.

By analogy with stress, the principal strain directions are found from the equations

$$\tan 2\theta = \frac{\gamma_{xy}}{\varepsilon_x - \varepsilon_y} \tag{3.29}$$

Similarly, the magnitudes of the principal strains are

$$\varepsilon_{1,2} = \frac{\varepsilon_x + \varepsilon_y}{2} \pm \sqrt{\left(\frac{\varepsilon_x - \varepsilon_y}{2}\right)^2 + \left(\frac{\gamma_{xy}}{2}\right)^2} \tag{3.30}$$

3.15 Equations of Compatibility for Strains

Expressions of compatibility have both mathematical and physical significance. From a mathematical point of view, they state that the displacements u, v,w are single valued and continuous functions. Physically, this means that the body must be pieced together.

It is to be noted that, if the three displacement components u, v and w are given as the continuous functions of the co-ordinates X, Y, Z then the six strain components can be uniquely obtained from the strain–displacement equations given below.

$$\varepsilon_x = \frac{\partial u}{\partial x}, \quad \varepsilon_y = \frac{\partial v}{\partial y}, \quad \varepsilon_z = \frac{\partial w}{\partial z}$$

$$\gamma_{xy} = \frac{\partial v}{\partial x} + \frac{\partial u}{\partial y}, \quad \gamma_{yz} = \frac{\partial w}{\partial y} + \frac{\partial v}{\partial z}, \quad \gamma_{zx} = \frac{\partial u}{\partial z} + \frac{\partial w}{\partial x}$$

However, if a strain field in terms of six strain components is arbitrarily specified in a body, then the displacement components cannot be uniquely determined easily. This is because; there are six equations for the three unknowns u, v and w. Therefore, if the values of the displacement components are to be single valued and continuous, then certain interrelationships between six strain components must exist. These interrelationships between the strains represent compatibility equations.

Now consider a body with a triangle PQR before straining as shown in Fig. 3.5a. The same triangle may take up one of the two possible positions as shown in Fig. 3.5b

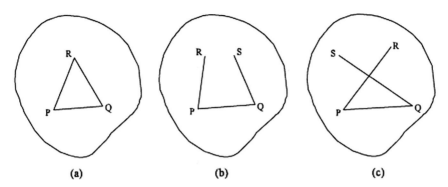

Fig. 3.5 Strain in a body

Fig. 3.6 Two-dimensional rotational transformation

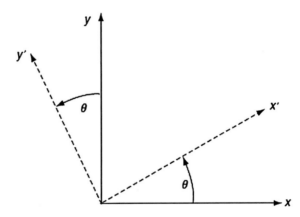

and in Fig. 3.5c after straining, if an arbitrary strain field is specified. Here, a gap or an overlapping may occur, unless the specified strain field obeys the necessary compatibility conditions.

Let us consider a set of strain data obtained from experiments on an externally loaded body for a possible state of strain as below.

$$\varepsilon_x = \frac{\partial u}{\partial x} = f(x, y) \tag{3.31}$$

$$\varepsilon_y = \frac{\partial v}{\partial y} = g(x, y) \tag{3.32}$$

$$\gamma_{xy} = \frac{\partial v}{\partial x} + \frac{\partial u}{\partial y} = h(x, y) \tag{3.33}$$

Let us take the derivatives of (a), (b) and (c) as:

$$\frac{\partial^2 f}{\partial y^2} = \frac{\partial^3 u}{\partial x \, \partial y^2} \tag{3.34}$$

$$\frac{\partial^2 g}{\partial x^2} = \frac{\partial^3 v}{\partial y \, \partial x^2} \tag{3.35}$$

$$\frac{\partial^2 h}{\partial x \, \partial y} = \frac{\partial^3 v}{\partial x^2 \, \partial y} + \frac{\partial^3 u}{\partial x \, \partial y^2} \tag{3.36}$$

From (3.34)–(3.36), we can write as:

$$\frac{\partial^2 h}{\partial x \, \partial y} = \frac{\partial^2 f}{\partial y^2} + \frac{\partial^2 g}{\partial x^2} \tag{3.37}$$

Hence, the experimental data must satisfy Eq. (3.37) in order to get consistent displacement. Equation (3.37) is known as equation of compatibility for strains. Hence, from (3.31)–(3.33), we can write.

$$\frac{\partial^2 \varepsilon_x}{\partial y^2} + \frac{\partial^2 \varepsilon_y}{\partial x^2} = \frac{\partial^2 \gamma_{xy}}{\partial x \, \partial y} \tag{3.38}$$

Similarly, we can get

$$\frac{\partial^2 \varepsilon_y}{\partial z^2} + \frac{\partial^2 \varepsilon_z}{\partial y^2} = \frac{\partial^2 \gamma_{yz}}{\partial y \, \partial z} \tag{3.39}$$

$$\frac{\partial^2 \varepsilon_z}{\partial x^2} + \frac{\partial^2 \varepsilon_x}{\partial z^2} = \frac{\partial^2 \gamma_{zx}}{\partial x \, \partial z} \tag{3.40}$$

Now, take the mixed derivative of ε_x with respect to z and y, Hence,

$$\frac{\partial^2 \varepsilon_x}{\partial y \, \partial z} = \frac{\partial^3 u}{\partial x \, \partial y \, \partial z} \tag{3.41}$$

and taking the partial derivative of γ_{xy} with respect to z and x, we get

$$\frac{\partial^2 \gamma_{xy}}{\partial x \, \partial z} = \frac{\partial^3 u}{\partial x \, \partial y \, \partial z} + \frac{\partial^3 v}{\partial z \, \partial x^2} \tag{3.42}$$

Also, take the partial derivative of γ_{yz} with respect to x twice, we get

$$\frac{\partial^2 \gamma_{yz}}{\partial x^2} = \frac{\partial^3 w}{\partial x^2 \partial y} + \frac{\partial^3 v}{\partial x^2 \partial z} \tag{3.43}$$

And take the derivative of γ_{zx} with respect to y and x.
Thus,

$$\frac{\partial^2 \gamma_{zx}}{\partial x \, \partial y} = \frac{\partial^3 u}{\partial x \, \partial y \, \partial z} + \frac{\partial^3 w}{\partial x^2 \, \partial y} \tag{3.44}$$

Now, adding Eqs. (3.42) and (3.44) and subtracting Eq. (3.43), we get

$$-\left(\frac{\partial^2 \gamma_{yz}}{\partial x^2}\right) + \frac{\partial^2 \gamma_{xz}}{\partial x \partial y} + \frac{\partial^2 \gamma_{xy}}{\partial x \partial z} = \frac{2 \partial^3 u}{\partial x \partial y \partial z} \tag{3.45}$$

From the above, we get

$$\frac{2 \partial^2 \varepsilon_x}{\partial y \partial z} = \frac{\partial}{\partial x}\left[-\frac{\partial \gamma_{yz}}{\partial x} + \frac{\partial \gamma_{xz}}{\partial y} + \frac{\partial \gamma_{xy}}{\partial z}\right] \tag{3.46}$$

Similarly, we can get

$$\frac{2 \partial^2 \varepsilon_y}{\partial x \partial z} = \frac{\partial}{\partial y}\left[-\frac{\partial \gamma_{zx}}{\partial y} + \frac{\partial \gamma_{yx}}{\partial z} + \frac{\partial \gamma_{yz}}{\partial x}\right] \tag{3.47}$$

$$\frac{2 \partial^2 \varepsilon_z}{\partial x \partial y} = \frac{\partial}{\partial z}\left[-\frac{\partial \gamma_{xy}}{\partial z} + \frac{\partial \gamma_{yz}}{\partial x} + \frac{\partial \gamma_{zx}}{\partial y}\right] \tag{3.48}$$

Thus, for a three-dimensional situation, the following are the six equations of compatibility of strain.

$$\begin{aligned}
\frac{\partial^2 \varepsilon_x}{\partial y^2} + \frac{\partial^2 \varepsilon_y}{\partial x^2} &= \frac{\partial^2 \gamma_{xy}}{\partial x \partial y} \\
\frac{\partial^2 \varepsilon_y}{\partial z^2} + \frac{\partial^2 \varepsilon_z}{\partial y^2} &= \frac{\partial^2 \gamma_{yz}}{\partial y \partial z} \\
\frac{\partial^2 \varepsilon_z}{\partial x^2} + \frac{\partial^2 \varepsilon_x}{\partial z^2} &= \frac{\partial^2 \gamma_{zx}}{\partial z \partial x} \\
\frac{2 \partial^2 \varepsilon_x}{\partial y \partial z} &= \frac{\partial}{\partial x}\left(-\frac{\partial \gamma_{yz}}{\partial x} + \frac{\partial \gamma_{xz}}{\partial y} + \frac{\partial \gamma_{xy}}{\partial z}\right) \\
\frac{2 \partial^2 \varepsilon_y}{\partial z \partial x} &= \frac{\partial}{\partial y}\left(\frac{\partial \gamma_{yz}}{\partial x} - \frac{\partial \gamma_{zx}}{\partial y} + \frac{\partial \gamma_{xy}}{\partial z}\right) \\
\frac{2 \partial^2 \varepsilon_z}{\partial x \partial y} &= \frac{\partial}{\partial z}\left(\frac{\partial \gamma_{yz}}{\partial x} + \frac{\partial \gamma_{zx}}{\partial y} - \frac{\partial \gamma_{xy}}{\partial z}\right)
\end{aligned} \tag{3.49}$$

3.16   Measurement of Surface Strains—Strain Rosettes

3.16.1   Strain Rosettes

Whenever a material is subjected to plane stress, it is desirable to obtain the stresses by direct measurement. As the stresses cannot be measured directly, it is essential to measure the strains or deformations that take place in the material during loading. These strains or deformations are measured with sensitive strain gauges attached to the surface of the body before it is loaded so that these gauges can record the amount of strain that takes place during loading.

It is more accurate and easier to measure in the neighbourhood of a chosen point on the surface of the body, the linear strains in different directions and then computes from these measurements the magnitudes and directions of the principal strains. Such a group of strain gauges is called a "strain rosette".

3.16.2   Strain Transformation Laws

If the components of strain at a point in a body are represented as ε_x, ε_y and γ_{xy} with reference to the rectangular co-ordinate axes X and Y, then the strain components with reference to a set of axes inclined at an angle θ with axis X (Fig. 3.6) can be expressed as.

$$\varepsilon_\theta = \left(\frac{\varepsilon_x + \varepsilon_y}{2} \right) + \left(\frac{\varepsilon_x - \varepsilon_y}{2} \right) \cos 2\theta + \frac{\gamma_{xy}}{2} \sin 2\theta \qquad (3.50)$$

$$\gamma_\theta = \left(\varepsilon_y - \varepsilon_x \right) \sin 2\theta + \gamma_{xy} \cos 2\theta \qquad (3.51)$$

and the principal strains are given by

$$\varepsilon_{\max} \text{ or } \varepsilon_{\min} = \left(\frac{\varepsilon_x + \varepsilon_y}{2} \right) \pm \frac{1}{2} \sqrt{\left(\varepsilon_x - \varepsilon_y \right)^2 + \gamma_{xy}^2} \qquad (3.52)$$

The direction of the principal strains are defined by the angle θ as

$$\tan 2\theta = \left(\frac{\gamma_{xy}}{\varepsilon_x - \varepsilon_y} \right) \qquad (3.53)$$

Also, the maximum shear strain at the point is given by following relation.

$$\gamma_{\max} = \sqrt{\left(\varepsilon_x - \varepsilon_y \right)^2 + \gamma_{xy}^2} \qquad (3.54)$$

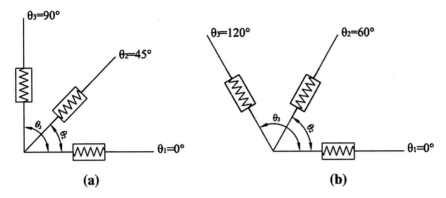

Fig. 3.7 Strain rosettes

3.16.3   *Measurement of Strains Using Rosettes*

In a rectangular rosette (also called 45° rosette) as shown in Fig. 3.7a, the strains are measured at angles denoted by $\theta_1 = 0°$, $\theta_2 = 45°$ and $\theta_3 = 90°$. In an equiangular rosette (also called delta rosette) as in Fig. 3.7b, the strains are measured at angles denoted by $\theta_1 = 0°$, $\theta_2 = 60°$, $\theta_3 = 120°$.

Let ε_{θ_1}, ε_{θ_2} and ε_{θ_3} are the strains measured at three different angles θ_1, θ_2 and θ_3, respectively. Now, using the transformation laws, we can write the three simultaneous equations as follows:

$$\varepsilon_{\theta_1} = \left(\frac{\varepsilon_x + \varepsilon_y}{2}\right) + \left(\frac{\varepsilon_x - \varepsilon_y}{2}\right)\cos 2\theta_1 + \frac{\gamma_{xy}}{2}\sin 2\theta_1 \qquad (3.55)$$

$$\varepsilon_{\theta_2} = \left(\frac{\varepsilon_x + \varepsilon_y}{2}\right) + \left(\frac{\varepsilon_x - \varepsilon_y}{2}\right)\cos 2\theta_2 + \frac{\gamma_{xy}}{2}\sin 2\theta_2 \qquad (3.56)$$

$$\varepsilon_{\theta_3} = \left(\frac{\varepsilon_x + \varepsilon_y}{2}\right) + \left(\frac{\varepsilon_x - \varepsilon_y}{2}\right)\cos 2\theta_3 + \frac{\gamma_{xy}}{2}\sin 2\theta_3 \qquad (3.57)$$

For a rectangular rosette,

$$\theta_1 = 0, \quad \theta_2 = 45° \text{ and } \theta_3 = 90°$$

Substituting the above in Eqs. (3.55)–(3.57), we get

$$\varepsilon_{0°} = \left(\frac{\varepsilon_x + \varepsilon_y}{2}\right) + \left(\frac{\varepsilon_x - \varepsilon_y}{2}\right) + 0$$

$$= \frac{1}{2}\left(\varepsilon_x + \varepsilon_y + \varepsilon_x - \varepsilon_y\right)$$

$$\therefore \varepsilon_{0^\circ} = \varepsilon_x$$

$$\varepsilon_{45} = \left(\frac{\varepsilon_x + \varepsilon_y}{2}\right) + \left(\frac{\varepsilon_x - \varepsilon_y}{2}\right)(0) + \frac{\gamma_{xy}}{2}$$
$$= \frac{1}{2}(\varepsilon_x + \varepsilon_y + \gamma_{xy})$$

$$\text{or } 2\varepsilon_{45} = \varepsilon_x + \varepsilon_y + \gamma_{xy}$$

$$\therefore \gamma_{xy} = 2\varepsilon_{45} - (\varepsilon_x + \varepsilon_y)$$

Also,

$$\varepsilon_{90} = \left(\frac{\varepsilon_x + \varepsilon_y}{2}\right) + \left(\frac{\varepsilon_x - \varepsilon_y}{2}\right)(\cos 180^\circ) + \frac{\gamma_{xy}}{2}(\sin 180^\circ)$$
$$= \left(\frac{\varepsilon_x + \varepsilon_y}{2}\right) - \left(\frac{\varepsilon_x - \varepsilon_y}{2}\right)$$
$$= \frac{1}{2}(\varepsilon_x + \varepsilon_y - \varepsilon_x + \varepsilon_y)$$
$$\therefore \varepsilon_{90} = \varepsilon_y$$

Therefore, the components of strain are given by

$$\varepsilon_x = \varepsilon_0, \quad \varepsilon_y = \varepsilon_{90^\circ} \quad \text{and} \quad \gamma_{xy} = 2\varepsilon_{45} - (\varepsilon_0 + \varepsilon_{90})$$

For an equiangular rosette,

$$\theta_1 = 0, \quad \theta_2 = 60^\circ, \quad \theta_3 = 120^\circ$$

Substituting the above values in (3.55)–(3.57), we get

$$\varepsilon_x = \varepsilon_0, \quad \varepsilon_y = \frac{1}{3}(2\varepsilon_{60} + 2\varepsilon_{120} - \varepsilon_0)$$

and $\gamma_{xy} = \frac{2}{\sqrt{3}}(\varepsilon_{60} - \varepsilon_{120})$.

Therefore, using the values of $\varepsilon_x, \varepsilon_y$ and γ_{xy}, the principal strains $\varepsilon_{\max}$ and $\varepsilon_{\min}$ can be computed. Further, the directions of these principal strains can be determined by Eq. (3.58).

$$\tan 2\theta = \left(\frac{\gamma_{xy}}{\varepsilon_x - \varepsilon_y}\right) \tag{3.58}$$

3.17 Numerical Examples

Example 3.1 A sheet of metal is deformed uniformly in its own plane that the strain components related to a set of axes xy are.

$$\varepsilon_x = -200 \times 10^{-6}$$
$$\varepsilon_y = 1000 \times 10^{-6}$$
$$\gamma_{xx} = 900 \times 10^{-6}$$

(a) Find the strain components associated with a set of axes $x'y'$ inclined at an angle of 30° clockwise to the xy set as shown in Fig. 3.8. Also, find the principal strains and the direction of the axes on which they act.

Solution: (a)

The transformation equations for strains similar to that for stresses can be written as below:

$$\varepsilon_{x'} = \frac{\varepsilon_x + \varepsilon_y}{2} + \frac{\varepsilon_x - \varepsilon_y}{2} \cos 2\theta + \frac{\gamma_{xy}}{2} \sin 2\theta$$
$$\varepsilon_{y'} = \frac{\varepsilon_x + \varepsilon_y}{2} - \frac{\varepsilon_x - \varepsilon_y}{2} \cos 2\theta - \frac{\gamma_{xy}}{2} \sin 2\theta$$
$$\frac{\gamma_{x'y'}}{2} = -\left(\frac{\varepsilon_x - \varepsilon_y}{2}\right) \sin 2\theta + \frac{\gamma_{xy}}{2} \cos 2\theta$$

Using Eq. (3.19), we find

$$2\theta = \tan^{-1}\left(\frac{450}{600}\right) = 36.8°$$

Radius of Mohr's $= R = \sqrt{(600)^2 + (450)^2} = 750.$
Therefore,

Fig. 3.8 Transformed co-ordinate system

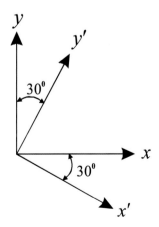

$$\varepsilon_{x'} = \left(400 \times 10^{-6}\right) - \left(750 \times 10^{-6}\right) \cos\left(60^0 - 36.8^0\right)$$
$$= -290 \times 10^{-6}$$

$$\varepsilon_{y'} = \left(400 \times 10^{-6}\right) + \left(750 \times 10^{-6}\right) \cos\left(60^0 - 36.8^0\right)$$
$$= 1090 \times 10^{-6}$$

Because point x' lies above the ε-axis and point y' below ε-axis, the shear strain $\gamma_{x'y'}$ is negative.

Therefore,

$$\frac{\gamma_{x'y'}}{2} = -\left(750 \times 10^{-6}\right) \sin\left(60^0 - 36.8^0\right)$$
$$= -295 \times 10^{-6}$$

Hence, $\gamma_{x'y'} = -590 \times 10^{-6}$.

Solution: (b)

From the Mohr's circle of strain, the principal strains are (Fig. 3.9)

$$\varepsilon_1 = 1150 \times 10^{-6}$$

$$\varepsilon_2 = -350 \times 10^{-6}$$

The directions of the principal axes of strain are shown in Fig. 3.10.

Example 3.2 By means of strain rosette, the following strains were recorded during the test on a structural member.

$$\varepsilon_0 = -13 \times 10^{-6} \text{ mm/mm}, \quad \varepsilon_{45} = 7.5 \times 10^{-6} \text{ mm/mm}, \quad \varepsilon_{90} = 13 \times 10^{-6} \text{ mm/mm}$$

Determine (a) Magnitude of principal strains
(b) Orientation of principal planes.
Solution: (a) We have for a rectangular strain rosette the following:

$$\varepsilon_x = \varepsilon_0 \qquad \varepsilon_y = \varepsilon_{90} \qquad \gamma_{xy} = 2\varepsilon_{45} - \left(\varepsilon_0 + \varepsilon_{90}\right)$$

Substituting the values in the above relations, we get

$$\varepsilon_x = -13 \times 10^{-6} \qquad \varepsilon_y = 13 \times 10^{-6}$$

$$\gamma_{xy} = 2 \times 7.5 \times 10^{-6} - \left(-12 \times 10^{-6} + 13 \times 10^{-6}\right) \qquad \therefore \quad \gamma_{xy} = 15 \times 10^{-6}$$

The principal strains can be determined from the following relation.

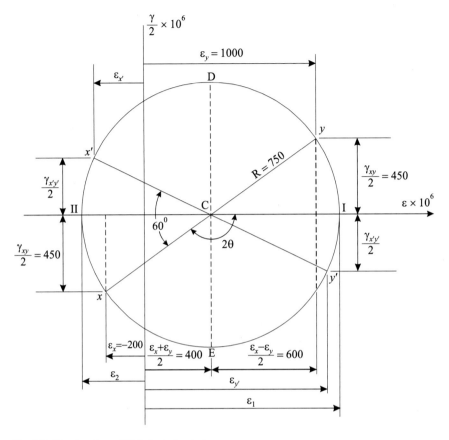

Fig. 3.9 Construction of Mohr's strain circle

Fig. 3.10 Directions of principal axes

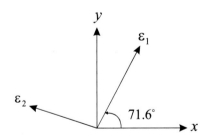

$$\varepsilon_{max} \text{ or } \varepsilon_{min} = \left(\tfrac{\varepsilon_x + \varepsilon_y}{2}\right) \pm \tfrac{1}{2}\sqrt{\left(\varepsilon_x - \varepsilon_y\right)^2 + \gamma_{xy}^2}$$

$$\therefore \varepsilon_{max} \text{ or } \varepsilon_{min} = \left(\tfrac{-13+13}{2}\right)10^{-6} \pm \tfrac{1}{2}\sqrt{\left[(-13 - 13)10^{-6}\right]^2 + \left(15 \times 10^{-6}\right)^2}$$

$$\therefore \varepsilon_{max} \text{ or } \varepsilon_{min} = \pm 15 \times 10^{-6}$$

Hence, $\varepsilon_{max} = 15 \times 10^{-6}$ and $\varepsilon_{min} = -15 \times 10^{-6}$.

Fig 3.11 Rotation of
coordinate axes

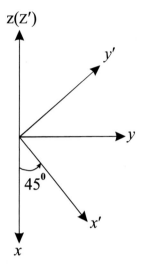

(b) The orientation of the principal strains can be obtained from the following relation

$$\tan 2\theta = \frac{\gamma_{xy}}{\left(\varepsilon_x - \varepsilon_y\right)}$$

$$= \frac{15 \times 10^{-6}}{(-13 - 13)10^{-6}}$$

$$\tan 2\theta = -0.577$$

$$\therefore 2\theta = 150°$$

$$\therefore \theta = 75°.$$

Hence, the directions of the principal planes are $\theta_1 = 75°$ and $\theta_2 = 165°$.

Example 3.3 Data taken from a 45° strain rosette reads as follows:

$$\varepsilon_0 = 750\,\mu\text{m/m}$$
$$\varepsilon_{45} = -110\,\mu\text{m/m}$$
$$\varepsilon_{90} = 210\,\mu\text{m/m}$$

Find the magnitudes and directions of principal strains.

Solution: Given

$$\varepsilon_0 = 750 \times 10^{-6}$$
$$\varepsilon_{45} = -110 \times 10^{-6}$$
$$\varepsilon_{90} = 210 \times 10^{-6}$$

Now, for a rectangular rosette,

$$\varepsilon_x = \varepsilon_0 = 750 \times 10^{-6}$$
$$\varepsilon_y = \varepsilon_{90} = 210 \times 10^{-6}$$
$$\gamma_{xy} = 2\varepsilon_{45} - (\varepsilon_0 + \varepsilon_{90})$$
$$= 2(-110 \times 10^{-6}) - (750 \times 10^{-6} + 210 \times 10^{-6})$$
$$\gamma_{xy} = -1180 \times 10^{-6}$$

$\therefore$ The magnitudes of principal strains are

$$\varepsilon_{\max} \text{ or } \varepsilon_{\min} = \left(\frac{\varepsilon_x + \varepsilon_y}{2}\right) \pm \frac{1}{2}\sqrt{(\varepsilon_x - \varepsilon_y)^2 + \gamma_{xy}^2}$$

i.e., $\varepsilon_{\max}$ or $\varepsilon_{\min} = \left(\dfrac{750 + 210}{2}\right)10^{-6} \pm \dfrac{1}{2}\sqrt{[(750-210)10^{-6}]^2 + [(-1180)10^{-6}]^2}$

$$= 480 \times 10^{-6} \pm \frac{1}{2}(1297.7)10^{-6}$$
$$= 480 \times 10^{-6} \pm 648.85 \times 10^{-6}$$

$$\therefore \varepsilon_{\max} = \varepsilon_1 = 1128.85 \times 10^{-6}$$

$$\varepsilon_{\min} = \varepsilon_2 = -168.85 \times 10^{-6}$$

The directions of the principal strains are given by the relation

$$\tan 2\theta = \frac{\gamma_{xy}}{(\varepsilon_x - \varepsilon_y)}$$

$$\therefore \tan 2\theta = \frac{-1180 \times 10^{-6}}{(750 - 210)10^{-6}} = -2.185$$

$$\therefore 2\theta = 114.6^0$$

$$\therefore \theta_1 = 57.3^0 \text{ and } \theta_2 = 147.3^0$$

Example 3.4 If the displacement field in a body is specified as $u = (x^2 + 3) \, 10^{-3}$, $v = 3y^2z \times 10^{-3}$ and $w = (x + 3z) \times 10^{-3}$, determine the strain components at a point whose co-ordinates are $(1, 2, 3)$.

Solution: From Eq. (3.3), we have

$$\varepsilon_x = \frac{\partial u}{\partial x} = 2x \times 10^{-3},$$

$$\varepsilon_y = \frac{\partial v}{\partial y} = 6yz \times 10^{-3},$$

$$\varepsilon_z = \frac{\partial w}{\partial z} = 3 \times 10^{-3}$$

$$\gamma_{xy} = \left[\frac{\partial}{\partial y}(x^2 + 3) \times 10^{-3} + \frac{\partial}{\partial x}(3y^2 z) \times 10^{-3} \right]$$

$$\gamma_{xy} = 0$$

$$\gamma_{yz} = \left[\frac{\partial}{\partial z}(3y^2 z \times 10^{-3}) + \frac{\partial}{\partial y}(x + 3z) \times 10^{-3} \right]$$

$$\gamma_{yz} = 3y^2 \times 10^{-3}$$

and $\gamma_{zx} = \left[\frac{\partial}{\partial x}(x + 3z)10^{-3} + \frac{\partial}{\partial z}(x^2 + 3)10^{-3} \right]$

$$\gamma_{zx} = 1 \times 10^{-3}$$

Therefore, at point (1, 2, 3), we get

$$\varepsilon_x = 2 \times 10^{-3}, \ \varepsilon_y = 6 \times 2 \times 3 \times 10^{-3} = 36 \times 10^{-3}, \ \varepsilon_z = 3 \times 10^{-3},$$
$$\gamma_{xy} = 0, \ \gamma_{yz} = 12 \times 10^{-3}, \ \gamma_{zx} = 1 \times 10^{-3}$$

Example 3.5 The strain components at a point with respect to $x\ y\ z$ co-ordinate system are

$$\varepsilon_x = 0.10, \ \varepsilon_y = 0.20, \ \varepsilon_z = 0.30, \ \gamma_{xy} = \gamma_{yz} = \gamma_{xz} = 0.160$$

If the co-ordinate axes are rotated about the z-axis through $45°$ in the anticlockwise direction (Fig. 3.11), determine the new strain components.
Solution: Direction cosines
Here,

$$l_1 = \frac{1}{\sqrt{2}}, \quad m_1 = \frac{1}{\sqrt{2}}, \quad n_1 = 0$$
$$l_2 = -\frac{1}{\sqrt{2}}, \quad m_2 = \frac{1}{\sqrt{2}}, \quad n_2 = 0$$

See Fig. 3.11.

$$l_3 = 0, \quad m_3 = 0, \quad n_3 = 1$$

Now, we have,

$$[\varepsilon'] = [a][\varepsilon][a]^T$$

$$[a][\varepsilon] = \begin{bmatrix} \frac{1}{\sqrt{2}} & \frac{1}{\sqrt{2}} & 0 \\ -\frac{1}{\sqrt{2}} & \frac{1}{\sqrt{2}} & 0 \\ 0 & 0 & 1 \end{bmatrix} \begin{bmatrix} 0.1 & 0.08 & 0.08 \\ 0.08 & 0.2 & 0.08 \\ 0.08 & 0.08 & 0.3 \end{bmatrix}$$

$$= \begin{bmatrix} 0.127 & 0.198 & 0.113 \\ -0.014 & 0.085 & 0 \\ 0.08 & 0.08 & 0.3 \end{bmatrix}$$

$$[\varepsilon'] = \begin{bmatrix} 0.127 & 0.198 & 0.113 \\ -0.014 & 0.085 & 0 \\ 0.08 & 0.08 & 0.3 \end{bmatrix} \begin{bmatrix} \frac{1}{\sqrt{2}} & -\frac{1}{\sqrt{2}} & 0 \\ \frac{1}{\sqrt{2}} & \frac{1}{\sqrt{2}} & 0 \\ 0 & 0 & 1 \end{bmatrix}$$

$$[\varepsilon'] = \begin{bmatrix} 0.23 & 0.05 & 0.113 \\ 0.05 & 0.07 & 0 \\ 0.113 & 0.3 & 0.3 \end{bmatrix}$$

Therefore, the new strain components are

$$\varepsilon_x = 0.23, \quad \varepsilon_y = 0.07, \quad \varepsilon_z = 0.3$$

$$\frac{1}{2}\gamma_{xy} = 0.05 \text{ or } \gamma_{xy} = 0.05 \times 2 = 0.1$$

$$\gamma_{yz} = 0, \quad \gamma_{zx} = 0.113 \times 2 = 0.226$$

Example 3.6 The components of strain at a point in a body are as follows:

$$\varepsilon_x = 0.1, \quad \varepsilon_y = -0.05, \quad \varepsilon_z = 0.05, \quad \gamma_{xy} = 0.3, \quad \gamma_{yz} = 0.1, \quad \gamma_{xz} = -0.08$$

Determine the principal strains and the principal directions.

Solution: The strain tensor is given by

$$\varepsilon_{ij} = \begin{bmatrix} \varepsilon_x & \frac{\gamma_{xy}}{2} & \frac{\gamma_{xz}}{2} \\ \frac{\gamma_{xy}}{2} & \varepsilon_y & \frac{\gamma_{yz}}{2} \\ \frac{\gamma_{xz}}{2} & \frac{\gamma_{yz}}{2} & \varepsilon_z \end{bmatrix} = \begin{bmatrix} 0.1 & 0.15 & -0.04 \\ 0.15 & -0.05 & 0.05 \\ -0.04 & 0.05 & 0.05 \end{bmatrix}$$

The invariants of strain tensor are

$$J_1 = \varepsilon_x + \varepsilon_y + \varepsilon_z = 0.1 - 0.05 + 0.05 = 0.1$$

$$J_2 = \varepsilon_x \varepsilon_y + \varepsilon_y \varepsilon_z + \varepsilon_z \varepsilon_x - \frac{1}{4}\left(\gamma_{xy}^2 + \gamma_{yz}^2 + \gamma_{zx}^2\right)$$

$$= (0.1)\,(-0.05) + (-0.05)\,(0.05) + (0.05)\,(0.1)$$

$$- \frac{1}{4}\left[(0.3)^2 + (0.1)^2 + (-0.08)^2\right]$$

$$\therefore J_2 = -0.0291$$

$$J_3 = (0.1)(-0.05)(0.05) + \frac{1}{4}[(0.3)(0.1)(-0.08) - 0.1(0.1)^2$$

$$+ 0.05(0.08)^2 - 0.05(0.3)^2$$

$$J_3 = -0.002145$$

∴ The cubic equation is

$$\varepsilon^3 - 0.1\varepsilon^2 - 0.0291\varepsilon + 0.002145 = 0 \tag{i}$$

Now $\cos 3\theta = 4\cos^3\theta - 3\cos\theta$

$$\text{Or} \quad \cos^3\theta - \frac{3}{4}\cos\theta - \frac{1}{4}\cos 3\theta = 0 \tag{ii}$$

$$\varepsilon = r\cos\theta + \frac{J_1}{3}$$

Let

$$= r\cos\theta + \frac{0.1}{3} \quad .$$

$$\varepsilon = r\cos\theta + 0.033$$

∴ (i) can be written as

$(r\cos\theta + 0.033)^3 - 0.1(r\cos\theta + 0.033)^2 - 0.0291(r\cos\theta + 0.033) + 0.002145$
$= 0$

$(r\cos\theta + 0.033)(r\cos\theta + 0.033)^2 - 0.1(r\cos\theta + 0.033)^2 - 0.0291r\cos\theta$
$-0.00096 + 0.002145 = 0$

$(r\cos\theta + 0.033)\left(r^2\cos^2\theta + 0.067r\cos\theta + 0.00109\right)$
$-0.1\left(r^2\cos^2\theta + 0.067r\cos\theta + 0.00109\right) - 0.0291r\cos\theta - 0.00096 + 0.002145$
$= 0$

$r^3\cos^3\theta + 0.067r^2\cos^2\theta + 0.00109r\cos\theta + 0.033r^2\cos^2\theta + 0.0022r\cos\theta$
$+0.000036 - 0.1r^2\cos^2\theta - 0.0067r\cos\theta - 0.000109 - 0.0291r\cos\theta - 0.00096$
$+0.002145 = 0$

i.e., $r^3\cos^3\theta - 0.03251r\cos\theta - 0.00112 = 0$

$$\text{or} \quad \cos^3\theta - \frac{0.03251}{r^2}\cos\theta - \frac{0.00112}{r^3} = 0 \tag{iii}$$

Hence, Eqs. (ii) and (iii) are identical, if

$$\frac{0.03251}{r^2} = \frac{3}{4}$$

i.e., $r = \sqrt{\frac{4 \times 0.03251}{3}} = 0.2082$

and $\frac{\cos 3\theta}{4} = \frac{0.00112}{r^3}$

or $\cos 3\theta = \frac{4 \times 0.00112}{(0.2082)^3} = 0.496 \cong 0.5$

$$\therefore 3\theta = 60^0 \text{ or } \theta_1 = \frac{60}{3} = 20^0$$

$$\theta_2 = 100^0 \qquad \theta_3 = 140^0$$

$$\therefore \varepsilon_1 = r_1 \cos \theta_1 + \frac{J_1}{3}$$

$$= 0.2082 \cos 20° + \frac{0.1}{3}$$

$$\varepsilon_1 = 0.228$$

$$\varepsilon_2 = r_2 \cos \theta_2 + \frac{J_1}{3} = 0.2082 \cos 100° + \frac{0.1}{3} = -0.0031$$

$$\varepsilon_3 = r_3 \cos \theta_3 + \frac{J_1}{3} = 0.2082 \cos 140° + \frac{0.1}{3} = -0.126$$

To find principal directions

(a) Principal direction for ε_1

$$\begin{bmatrix} (0.1 - \varepsilon_1) & 0.15 & -0.04 \\ 0.15 & (-0.05 - \varepsilon_1) & 0.05 \\ -0.04 & 0.05 & (0.05 - \varepsilon_1) \end{bmatrix}$$

$$= \begin{bmatrix} (0.1 - 0.228) & 0.15 & -0.04 \\ 0.15 & (-0.05 - 0.228) & 0.05 \\ -0.04 & 0.05 & (0.05 - 0.228) \end{bmatrix}$$

$$= \begin{bmatrix} -0.128 & 0.15 & -0.04 \\ 0.15 & -0.278 & 0.05 \\ -0.04 & 0.05 & -0.178 \end{bmatrix}$$

Now, $A_1 = \begin{vmatrix} -0.278 & 0.05 \\ 0.05 & -0.178 \end{vmatrix} = (-0.278)(-0.178) - (0.05)(0.05)$

$$\therefore A_1 = 0.046984$$

$$B_1 = - \begin{vmatrix} 0.15 & 0.05 \\ -0.04 & -0.178 \end{vmatrix} = -[0.15 \times (-0.178) + (0.05)(0.04)]$$

$$\therefore B_1 = 0.0247$$

$$C_1 = \begin{vmatrix} 0.15 & -0.278 \\ -0.04 & 0.05 \end{vmatrix} = 0.15 \times 0.05 - 0.278 \times 0.04$$

$$\therefore C_1 = -0.00362$$

$$\sqrt{A_1^2 + B_1^2 + C_1^2} = \sqrt{(0.046984)^2 + (0.0247)^2 + (-0.00362)^2}$$
$$= 0.0532$$

$$\therefore l_1 = \frac{A_1}{\sqrt{A_1^2 + B_1^2 + C_1^2}} = \frac{0.046984}{0.0532} = 0.883$$

$$m_1 = \frac{B_1}{\sqrt{A_1^2 + B_1^2 + C_1^2}} = \frac{0.0247}{0.0532} = 0.464$$

$$n_1 = \frac{C_1}{\sqrt{A_1^2 + B_1^2 + C_1^2}} = \frac{-0.00362}{0.0532} = -0.068$$

Similarly, the principal directions for ε_2 can be determined as follows:

$$\begin{bmatrix} (0.1 + 0.0031) & 0.15 & -0.04 \\ 0.15 & (-0.05 + 0.0031) & 0.05 \\ -0.04 & 0.05 & (0.05 + 0.0031) \end{bmatrix}$$
$$= \begin{bmatrix} 0.1031 & 0.15 & -0.04 \\ 0.15 & -0.0469 & 0.05 \\ -0.04 & 0.05 & 0.0531 \end{bmatrix}$$

$$A_2 = \begin{vmatrix} -0.0469 & 0.05 \\ 0.05 & 0.0531 \end{vmatrix} = -0.00249 - 0.0025 = -0.00499$$

$$B_2 = \begin{vmatrix} 0.15 & 0.05 \\ -0.04 & 0.0531 \end{vmatrix} = -(0.007965 + 0.002) = -0.009965$$

$$C_2 = \begin{vmatrix} 0.15 & -0.0469 \\ -0.04 & 0.05 \end{vmatrix} = 0.0075 - 0.00188 = 0.00562$$

Now, $\sqrt{A_2^2 + B_2^2 + C_2^2} = \sqrt{(-0.00499)^2 + (-0.009965)^2 + (0.00562)^2} = 0.0125$

$$\therefore l_2 = \frac{A_2}{\sqrt{A_2^2 + B_2^2 + C_2^2}} = \frac{-0.00499}{0.0125} = -0.399$$

$$m_2 = \frac{B_2}{\sqrt{A_2^2 + B_2^2 + C_2^2}} = \frac{-0.009965}{0.0125} = -0.797$$

$$n_2 = \frac{C_2}{\sqrt{A_2^2 + B_2^2 + C_2^2}} = \frac{0.00562}{0.0125} = 0.450$$

And for $\varepsilon_3 = -0.126$

$$\begin{vmatrix} (0.1 + 0.126) & 0.15 & -0.04 \\ 0.15 & (-0.05 + 0.126) & 0.05 \\ -0.04 & 0.05 & (0.05 + 0.126) \end{vmatrix}$$

$$= \begin{vmatrix} 0.226 & 0.15 & -0.04 \\ 0.15 & 0.076 & 0.05 \\ -0.04 & 0.05 & 0.176 \end{vmatrix}$$

Now,

$$A_3 = \begin{vmatrix} 0.076 & 0.05 \\ 0.05 & 0.176 \end{vmatrix} = 0.0134 - 0.0025 = 0.0109$$

$$B_3 = - \begin{vmatrix} 0.15 & 0.05 \\ -0.04 & 0.176 \end{vmatrix} = -(0.0264 + 0.002) = -0.0284$$

$$C_3 = - \begin{vmatrix} 0.15 & 0.076 \\ -0.04 & 0.05 \end{vmatrix} = 0.0075 + 0.00304 = 0.01054.$$

Now, $\sqrt{A_3^2 + B_3^2 + C_3^2} = \sqrt{(0.0109)^2 + (-0.0284)^2 + (0.01054)^2} = 0.0322$

$$\therefore l_3 = \frac{A_3}{\sqrt{A_3^2 + B_3^2 + C_3^2}} = \frac{0.0109}{0.0322} = 0.338$$

$$m_3 = \frac{B_3}{\sqrt{A_3^2 + B_3^2 + C_3^2}} = \frac{-0.0284}{0.0322} = -0.882$$

$$n_3 = \frac{C_3}{\sqrt{A_3^2 + B_3^2 + C_3^2}} = \frac{0.01054}{0.0322} = 0.327$$

Example 3.7 The displacement components in a strained body are as follows. Determine the strain matrix at the point P (3, 2, −5).

$$u = 0.01xy + 0.02y^2, \quad v = 0.02x^2 + 0.01z^3y, \quad w = 0.01xy^2 + 0.05z^2$$

Solution:

$$\varepsilon_x = \frac{\partial u}{\partial x} = 0.01y$$
$$\varepsilon_y = \frac{\partial v}{\partial y} = 0.01z^3$$
$$\varepsilon_z = \frac{\partial w}{\partial z} = 0.1z$$
$$\gamma_{xy} = \frac{\partial v}{\partial x} + \frac{\partial u}{\partial y} = 0.04x + 0.01x + 0.04y$$
$$\gamma_{yz} = \frac{\partial w}{\partial y} + \frac{\partial v}{\partial z} = 0.02xy + 0.03z^2y$$
$$\gamma_{zx} = \frac{\partial u}{\partial z} + \frac{\partial w}{\partial x} = 0 + 0.01y^2$$

At point P (3, 2, −5), the strain components are

$$\varepsilon_x = 0.02, \quad \varepsilon_y = -1.25, \quad \varepsilon_z = -0.5$$
$$\gamma_{xy} = 0.23, \quad \gamma_{yz} = 1.62, \quad \gamma_{zx} = 0.04$$

Now, the strain tensor is given by

$$\varepsilon_{ij} = \begin{bmatrix} \varepsilon_x & \frac{1}{2}\gamma_{xy} & \frac{1}{2}\gamma_{xz} \\ \frac{1}{2}\gamma_{yx} & \varepsilon_y & \frac{1}{2}\gamma_{yz} \\ \frac{1}{2}\gamma_{zx} & \frac{1}{2}\gamma_{zy} & \varepsilon_z \end{bmatrix}$$

∴ Strain matrix becomes

$$\varepsilon_{ij} = \begin{bmatrix} 0.02 & 0.115 & 0.02 \\ 0.115 & -1.25 & 0.81 \\ 0.02 & 0.81 & -0.50 \end{bmatrix}$$

Example 3.8 The strain tensor at a point in a body is given by

$$\varepsilon_{ij} = \begin{bmatrix} 0.0001 & 0.0002 & 0.0005 \\ 0.0002 & 0.0003 & 0.0004 \\ 0.0005 & 0.0004 & 0.0005 \end{bmatrix}$$

Determine (a) octahedral normal and shearing strains. (b) Deviator and spherical strain tensors.

Solution: For the octahedral plane, the direction cosines are $l = m = n = \frac{1}{\sqrt{3}}$.

(a) Octahedral normal strain is given by

$$(\varepsilon_n)_{oct} = \varepsilon_x l^2 + \varepsilon_y m^2 + \varepsilon_z n^2 + 2(\varepsilon_{xy}lm + \varepsilon_{yz}mn + \varepsilon_{zx}nl)$$

Here, $\varepsilon_{xy} = \frac{1}{2}\gamma_{xy}$, $\quad \varepsilon_{yz} = \frac{1}{2}\gamma_{yz}$ and $\varepsilon_{zx} = \frac{1}{2}\gamma_{zx}$

$$\therefore \; (\varepsilon_n)_{oct} = 0.0001\left(\frac{1}{\sqrt{3}}\right)^2 + 0.0003\left(\frac{1}{\sqrt{3}}\right)^2 + 0.0005\left(\frac{1}{\sqrt{3}}\right)^2$$
$$+2\left[0.0002\left(\tfrac{1}{3}\right) + 0.0004\left(\tfrac{1}{3}\right) + 0.0005\left(\tfrac{1}{3}\right)\right]$$

$$\therefore \; (\varepsilon_n)_{oct} = 0.001$$

Octahedral shearing strain is given by

$$\gamma_{oct} = 2\sqrt{(\varepsilon_R)^2_{oct} - (\varepsilon_n)^2_{oct}}$$

where $(\varepsilon_R)_{oct}$ = Resultant strain on octahedral plane

$$\therefore \; (\varepsilon_R)_{oct} = \sqrt{\frac{1}{3}\left[\left(\varepsilon_x + \varepsilon_{xy} + \varepsilon_{xz}\right)^2 + \left(\varepsilon_{xy} + \varepsilon_y + \varepsilon_{yz}\right)^2 + \left(\varepsilon_{xz} + \varepsilon_{yz} + \varepsilon_y\right)^2\right]}$$
$$= \sqrt{\frac{1}{3}\left[(0.0001 + 0.0002 + 0.0005)^2 + (0.0002 + 0.0003 + 0.0004)^2(0.0005 + 0.0004 + 0.0005)^2\right]}$$

$$\therefore \; (\varepsilon_R)_{oct} = 0.001066$$

$$\therefore \; \gamma_{oct} = 2\sqrt{(0.00106)^2 - (0.001)^2}$$

$$\therefore \; \gamma_{oct} = 0.000739$$

(b) Deviator and spherical strain tensors.

Here,

$$\text{mean strain} = \varepsilon_m = \frac{\varepsilon_x + \varepsilon_y + \varepsilon_z}{3}$$
$$= \frac{0.0001 + 0.0003 + 0.0005}{3}$$

$$\therefore \; \varepsilon_m = 0.0003$$

∴ Deviator strain tensor =

$$
\begin{bmatrix}
(0.0001 - 0.0003) & 0.0002 & 0.0005 \\
0.0002 & (0.0003 - 0.0003) & 0.0004 \\
0.0005 & 0.0004 & (0.0005 - 0.0003)
\end{bmatrix}.
$$

i.e., $E' =$
$$
\begin{bmatrix}
-0.0002 & 0.0002 & 0.0005 \\
0.0002 & 0 & 0.0004 \\
0.0005 & 0.0004 & 0.0002
\end{bmatrix}.
$$

and spherical strain tensor $= E'' =$
$$
\begin{bmatrix}
\varepsilon_m & 0 & 0 \\
0 & \varepsilon_m & 0 \\
0 & 0 & \varepsilon_m
\end{bmatrix}.
$$

i.e., $E'' =$
$$
\begin{bmatrix}
0.0003 & 0 & 0 \\
0 & 0.0003 & 0 \\
0 & 0 & 0.0003
\end{bmatrix}.
$$

Example 3.9 The components of strain at a point in a body are as follows:

$$
\varepsilon_x = c_1 z \left(x^2 + y^2 \right)
$$
$$
\varepsilon_y = x^2 z
$$
$$
\gamma_{xy} = 2c_2 xyz
$$

where c_1 and c_2 are constants. Check whether the strain field is compatible one?

Solution: For the compatibility condition of the strain field, the system of strains must satisfy the compatibility equations.

i.e., $\frac{\partial^2 \varepsilon_x}{\partial y^2} + \frac{\partial^2 \varepsilon_y}{\partial x^2} = \frac{\partial^2 \gamma_{xy}}{\partial x \partial y}$

Now, using the given strain field,

$$
\frac{\partial \varepsilon_x}{\partial y} = 2c_1 yz, \qquad \frac{\partial^2 \varepsilon_x}{\partial y^2} = 2c_1 z
$$

$$
\frac{\partial \varepsilon_y}{\partial x} = 2xz, \qquad \frac{\partial^2 \varepsilon_y}{\partial x^2} = 2z
$$

$$
\frac{\partial \gamma_{xy}}{\partial x} = 2c_2 yz, \qquad \frac{\partial^2 \gamma_{xy}}{\partial x \partial y} = 2c_2 z
$$

$$
\therefore \frac{\partial^2 \varepsilon_x}{\partial y^2} + \frac{\partial^2 \varepsilon_y}{\partial x^2} = 2c_1 z + 2z = 2z(1 + c_1) \quad \text{and} \quad \frac{\partial^2 \gamma_{xy}}{\partial x \partial y} = 2c_2 z
$$

Since $\frac{\partial^2 \varepsilon_x}{\partial y^2} + \frac{\partial^2 \varepsilon_y}{\partial x^2} \neq \frac{\partial^2 \gamma_{xy}}{\partial x \partial y}$, the strain field is not compatible.

Example 3.10 Under what conditions are the following expressions for the components of strain at a point compatible?

$$\varepsilon_x = 2axy^2 + by^2 + 2cxy$$
$$\varepsilon_y = ax^2 + bx$$
$$\gamma_{xy} = \alpha x^2 y + \beta xy + ax^2 + \eta y$$

Solution: For compatibility, the strain components must satisfy the compatibility equation.

$$\text{i.e.,} \quad \frac{\partial^2 \varepsilon_x}{\partial y^2} + \frac{\partial^2 \varepsilon_y}{\partial x^2} = \frac{\partial^2 \gamma_{xy}}{\partial x \partial y} \tag{i}$$

$$\text{or} \quad \frac{\partial^2 \varepsilon_x}{\partial y^2} + \frac{\partial^2 \varepsilon_y}{\partial x^2} - \frac{\partial^2 \gamma_{xy}}{\partial x \partial y} = 0 \tag{ii}$$

Now, $\varepsilon_x = 2axy^2 + by^2 + 2cxy$

$$\therefore \frac{\partial \varepsilon_x}{\partial y} = 4axy + 2by + 2cx$$

$$\frac{\partial^2 \varepsilon_x}{\partial y^2} = 4ax + 2b$$

$$\varepsilon_y = ax^2 + bx$$

$$\frac{\partial \varepsilon_y}{\partial x} = 2ax + b$$

$$\frac{\partial^2 \varepsilon_y}{\partial x^2} = 2a$$

$$\gamma_{xy} = \alpha x^2 y + \beta xy + ax^2 + \eta y$$

$$\frac{\partial \gamma_{xy}}{\partial x} = 2\alpha xy + \beta y + 2ax$$

$$\frac{\partial^2 \gamma_{xy}}{\partial x \partial y} = 2\alpha x + \beta$$

$\therefore$ (i) becomes

$$4ax + 2b + 2a = 2\alpha x + \beta$$

$$4ax + 2(a + b) = 2\alpha x + \beta$$

$$\therefore 4ax = 2\alpha x$$

or $\alpha = 2a$.

and $\beta = 2(a + b)$.

Example 3.11 For the given displacement field

$$u = c\left(x^2 + 2x\right)$$
$$w = 4cz^2$$
$$v = c\left(4x + 2y^2 + z\right)$$

where c is a very small constant, determine the strain at $(2, 1, 3)$, in the direction $0, -\frac{1}{\sqrt{2}}, \frac{1}{\sqrt{2}}$

Solution:

$$
\begin{aligned}
\varepsilon_x &= \tfrac{\partial u}{\partial x} = 2cx, & \gamma_{xy} &= \tfrac{\partial v}{\partial x} + \tfrac{\partial u}{\partial y} = 4c + 0 = 4c \\
\varepsilon_y &= \tfrac{\partial v}{\partial y} = 4cy, & \gamma_{yz} &= \tfrac{\partial w}{\partial y} + \tfrac{\partial v}{\partial z} = 0 + c = c \\
\varepsilon_z &= \tfrac{\partial w}{\partial z} = 8cz, & \gamma_{zx} &= \tfrac{\partial u}{\partial z} + \tfrac{\partial w}{\partial x} = 2c + 0 = 2c
\end{aligned}
$$

$\therefore$ At point $(2, 1, 3)$,

$$
\begin{aligned}
\varepsilon_x &= 2c \times 2 = 4c, & \gamma_{xy} &= 4c \\
\varepsilon_y &= 4c \times 1 = 4c, & \gamma_{yz} &= c \\
\varepsilon_z &= 8c \times 3 = 24c, & \gamma_{zx} &= 2c
\end{aligned}
$$

$\therefore$ The resultant strain in the direction $l = 0, \ m = -\frac{1}{\sqrt{2}}, \ n = \frac{1}{\sqrt{2}}$ is given by

$$\varepsilon_r = \varepsilon_x l^2 + \varepsilon_y m^2 + \varepsilon_z n^2 + \gamma_{xy} lm + \gamma_{yz} mn + \gamma_{zx} nl$$

$$= 0 + 4c\left(-\frac{1}{\sqrt{2}}\right)^2 + 24c\left(\frac{1}{\sqrt{2}}\right)^2 + 4c(0) + c\left(-\frac{1}{\sqrt{2}}\right)\left(\frac{1}{\sqrt{2}}\right) + 2c(0)$$

$$\therefore \varepsilon_r = 13.5\,c$$

Example 3.12 The strain components at a point are given by

$$\varepsilon_x = 0.01, \quad \varepsilon_y = -0.02, \quad \varepsilon_z = 0.03, \quad \gamma_{xy} = 0.015, \quad \gamma_{yz} = 0.02, \quad \gamma_{xz} = -0.01$$

Determine the normal and shearing strains on the octahedral plane.

Solution: An octahedral plane is one which is inclined equally to the three principal co-ordinates. Its direction cosines are $\frac{1}{\sqrt{3}}, \frac{1}{\sqrt{3}}, \frac{1}{\sqrt{3}}$.

Now, the normal strain on the octahedral plane is

$$(\varepsilon_n)_{oct} = \varepsilon_x l^2 + \varepsilon_y m^2 + \varepsilon_z n^2 + \gamma_{xy} lm + \gamma_{yz} mn + \gamma_{zx} nl$$

$$= \frac{1}{3}[0.01 - 0.02 + 0.03 + 0.015 + 0.02 - 0.01]$$

$$\therefore (\varepsilon_n)_{oct} = 0.015$$

The strain tensor can be written as

$$\begin{pmatrix} \varepsilon_x & \varepsilon_{xy} & \varepsilon_{xz} \\ \varepsilon_{xy} & \varepsilon_y & \varepsilon_{yz} \\ \varepsilon_{xz} & \varepsilon_{yz} & \varepsilon_z \end{pmatrix} = \begin{pmatrix} 0.01 & \frac{0.015}{2} & -\frac{0.01}{2} \\ \frac{0.015}{2} & -0.02 & \frac{0.02}{2} \\ -\frac{0.01}{2} & \frac{0.02}{2} & 0.03 \end{pmatrix} = \begin{pmatrix} 0.01 & 0.0075 & -0.005 \\ 0.0075 & -0.02 & 0.01 \\ -0.005 & 0.01 & 0.03 \end{pmatrix}$$

Now, the resultant strain on the octahedral plane is given by

$$(\varepsilon_R)_{oct} = \sqrt{\frac{1}{3}\left\{ (\varepsilon_x + \varepsilon_{xy} + \varepsilon_{xz})^2 + (\varepsilon_{xy} + \varepsilon_y + \varepsilon_{yz})^2 + (\varepsilon_{xz} + \varepsilon_{yz} + \varepsilon_z)^2 \right\}}$$

$$= \sqrt{\frac{1}{3}\left\{ (0.01 + 0.0075 - 0.005)^2 + (0.0075 - 0.02 + 0.01)^2 + (-0.005 + 0.01 + 0.03)^2 \right\}}$$

$$= \sqrt{0.0004625}$$

$$\therefore (\varepsilon_R)_{oct} = 0.0215$$

and octahedral shearing strain is given by

$$(\varepsilon_S)_{oct} = 2\sqrt{(\varepsilon_R)^2 - (\varepsilon_n)^2} = 2\sqrt{(0.0215)^2 - (0.015)^2}$$

$$\therefore (\varepsilon_S)_{oct} = 0.031$$

Example 3.13 The displacement field is given by

$$u = K\left(x^2 + 2z\right), \quad v = K\left(4x + 2y^2 + z\right), \quad w = 4Kz^2$$

where K is a very small constant. What are the strains at $(2, 2, 3)$ in directions

(a) $l = 0 \; m = \frac{1}{\sqrt{2}} \; n = \frac{1}{\sqrt{2}}$, (b) $l = 1 \; m = n = 0$, (c) $l = 0.6 \; m = 0 \; n = 0.8$

Solution:

$$\varepsilon_x = \frac{\partial u}{\partial x} = 2Kx, \quad \varepsilon_y = \frac{\partial v}{\partial y} = 4Ky, \quad \varepsilon_z = \frac{\partial w}{\partial z} = 8Kz$$

$$\gamma_{xy} = \frac{\partial v}{\partial x} + \frac{\partial u}{\partial y} = 4K + 0 = 4K$$

$$\gamma_{yz} = \frac{\partial w}{\partial y} + \frac{\partial v}{\partial z} = 0 + K = K$$

$$\gamma_{zx} = \frac{\partial u}{\partial z} + \frac{\partial w}{\partial x} = 2K + 0 = 2K$$

$\therefore$ At point (2, 2, 3),

$$\varepsilon_x = 4K, \qquad \varepsilon_y = 8K, \qquad \varepsilon_z = 24K$$
$$\gamma_{xy} = 4K, \qquad \gamma_{yz} = K, \qquad \gamma_{zx} = 2K$$

Now, the strain in any direction is given by
$$\varepsilon_r = \varepsilon_x l^2 + \varepsilon_y m^2 + \varepsilon_z n^2 + \gamma_{xy} lm + \gamma_{yz} mn + \gamma_{zx} nl \text{ (i)}$$
Case (a)
Substituting the values in expression (i), we get

$$\varepsilon_r = 4K(0) + 8K\left(\frac{1}{\sqrt{2}}\right)^2 + 24K\left(\frac{1}{\sqrt{2}}\right)^2 + 4K(0) + K\left(\frac{1}{\sqrt{2}}\right)\left(\frac{1}{\sqrt{2}}\right) + 2K(0)$$
$$\therefore \varepsilon_r = 4K + 12K + \frac{1}{2}K$$
$$\therefore \varepsilon_r = 16.5K$$

Case (b)

$$\varepsilon_r = 4K(1)^2 + 8K(0) + 24(0) + 4K(0) + K(0) + 2K(0)$$
$$\therefore \varepsilon_r = 4K$$

Case (c)

$$\varepsilon_r = 4K(0.6)^2 + 8K(0) + 24(0.8)^2 + 4K(0) + K(0) + 2K(0.8)(0.6)$$
$$\therefore \varepsilon_r = 17.76K$$

3.18   Exercises

1. Explain strain tensor.
2. Derive the strain–displacement relation at any point in an elastic body.
3. The displacement at a point (x, y) is as given below

$$u = 5x^4 + 3x^2 y^2 + x + y$$
$$v = y^3 + 2xy + 4$$

Compute the values of normal and shearing strains at a point $(3, -2)$ and verify whether compatibility exists or not?

4. Determine the strain components at point $(1, 2, 3)$ for the following displacement field.

$$u = 8x^2 + 2y + 6z + 10$$

$$v = 2x^3 + 6x^2 + y^2 + z + 5$$
$$w = x^3 + 3y^3 + 8xy + 4$$

5. Derive the compatibility equation in terms of strain and displacements.
6. At a point in a stressed material, the stresses acting are: $\sigma_x = 300\,\text{N/mm}^2$, $\sigma_y = 250\,\text{N/mm}^2$ and $\sigma_z = 220\,\text{N/mm}^2$. If $\gamma = 0.3$, calculate the volumetric strain.
7. In a steel bar subjected to three-dimensional stress system, the elongations measured in the three principal directions over a length of 100 cm were found to be 1.8 mm, 1.2 mm and 0.6 mm, respectively, along the x, y and z axes. Calculate the volumetric strain and new volume of the material.
8. The displacement components in a strained body are:

$$u = 0.02xy + 0.03y^2$$
$$v = 0.03x^2 + 0.02z^3y$$
$$w = 0.02xy^2 + 0.06z^2$$

Determine the strain matrix at the point $(3, 2, -5)$.
9. The strain components at a point with respect to xyz co-ordinate system are:

$$\varepsilon_x = 0.01 \qquad \varepsilon_y = 0.02 \qquad \varepsilon_z = 0.03$$
$$\gamma_{xy} = \gamma_{yz} = \gamma_{xz} = 0.016$$

If the co-ordinate axes are rotated about z-axis through $45°$ in the anticlockwise direction, determine the new strain component.
10. The components of strain tensor at a given point are given by the following array of terms:

$$\varepsilon_{ij} = \begin{bmatrix} 0.01 & 0.02 & 0.05 \\ 0.02 & 0.03 & 0.04 \\ 0.05 & 0.04 & 0.05 \end{bmatrix}$$

Determine

(a) Octahedral normal and shearing strains.
(b) Deviator and spherical strain tensors.

11. The displacement field components at a point are given by

$$u = -0.01y^2 + 0.15xyz$$
$$v = 0.02x^2y + 0.03x^2z$$
$$w = 0.15xyz - 0.01x^2yz$$

Determine the strain tensor at the point $(2, -1, 3)$.
12. At a point in a body, the components of strain are

$$\varepsilon_x = -0.000832 \qquad \varepsilon_y = -0.000832 \qquad \varepsilon_z = 0.001664$$
$$\gamma_{xy} = 0 \qquad \gamma_{yz} = 0.00145 \qquad \gamma_{xz} = 0$$

Find the principal strains.

13. The components of strain at a point in a body are

$$\varepsilon_x = 0.01 \qquad \varepsilon_y = -0.05 \qquad \varepsilon_z = 0.05$$
$$\gamma_{xy} = 0.03 \qquad \gamma_{yz} = 0.01 \qquad \gamma_{xz} = 0.008$$

Find the principal strains.

14. At a point in a material, the state of strain is represented by

$$\varepsilon_x = 0.00233 \qquad \varepsilon_{xy} = -0.00152$$
$$\varepsilon_y = 0.00091 \qquad \varepsilon_{yz} = 0.00085$$
$$\varepsilon_z = 0.00125 \qquad \varepsilon_{zx} = 0.00110$$

Find the direction cosines of the principal strains.

15. The principal strains at a point are given by

$$\varepsilon_1 = 2 \times 10^{-3} \qquad \varepsilon_2 = -3 \times 10^{-3} \qquad \varepsilon_3 = -4 \times 10^{-3}$$

Calculate the octahedral normal and shearing strains.

16. The strain components at a point are given by

$$\varepsilon_x = 10xy + 12z; \quad \gamma_{xy} = 4xy^2$$
$$\varepsilon_y = 6xy^2 + 2yz; \quad \gamma_{yz} = 2yz^2$$
$$\varepsilon_z = 2x^2z + 2y; \quad \gamma_{xz} = 2xz^2$$

Verify whether the compatibility equations are satisfied or not at the point (1, −1, 2).

17. For the given displacement field,

$$u = c(x^2 + 2z), \quad v = c(4x + 2y^2 + z), \quad w = 4cz^2,$$

where c is a very small constant, determine the strain at (2, 1, 3) in the direction.

$$0, \quad -\frac{1}{\sqrt{2}}, \quad \frac{1}{\sqrt{2}}.$$

18. A state of plane strain in a steel plate is defined by the following data

$$\varepsilon_x = 0.00050 \qquad \varepsilon_y = 0.00014 \qquad \varepsilon_z = 0.00036$$

Construct a Mohr's circle and find the magnitudes and directions of principal strains.

19. The following strains were measured in a structure during the test by means of strain gauges

$$\varepsilon_0 = 650 \times 10^{-6}$$
$$\varepsilon_{60} = -200 \times 10^{-6}$$
$$\varepsilon_{120} = 250 \times 10^{-6}.$$

Determine the following

(a) Magnitude of principal strains
(b) Orientation of principal planes.

20. Using an equiangular strain rosette, the following strains were measured at a point in a material.

$$\varepsilon_0 = 600\,\mu\text{m/ms}, \quad \varepsilon_{60} = -200\,\mu\text{m/m}, \quad \varepsilon_{120} = 300\,\mu\text{m/m}.$$

Calculate the magnitudes and directions of principal strains.

Chapter 4
Stress–Strain Relations

4.1 Introduction

In the previous chapters, the state of stress at a point was defined in terms of six components of stress, and in addition, three equilibrium equations were developed to relate the internal stresses and the applied forces. These relationships were independent of the deformations (strains) and the material behaviour. Hence, these equations are applicable to all types of materials.

Also, the state of strain at a point was defined in terms of six components of strain. These six strain–displacement relations and compatibility equations were derived in order to relate uniquely the strains and the displacements at a point. These equations were also independent of the stresses and the material behaviour and hence are applicable to all materials.

Irrespective of the independent nature of the equilibrium equations and strain–displacement relations, in practice, it is essential to study the general behaviour of materials under applied loads including these relations. This becomes necessary due to the application of a load, stresses, deformations and hence strains will develop in a body. Therefore in a general three-dimensional system, there will be 15 unknowns, namely 3 displacements, 6 strains and 6 stresses. In order to determine these 15 unknowns, we have only 9 equations such as 3 equilibrium equations and 6 strain–displacement equations. It is important to note that the compatibility conditions as such cannot be used to determine either the displacements or strains. Hence, the additional six equations have to be based on the relationships between six stresses and six strains. These equations are known as "Constitutive equations" because they describe the macroscopic behaviour of a material based on its internal constitution.

T. G. Sitharam and L. Govindaraju, *Theory of Elasticity*,
https://doi.org/10.1007/978-981-33-4650-5_4

4.2   Linear Elasticity—Generalized Hooke's Law

There is a unique relationship between stress and strain defined by Hooke's law, which is independent of time and loading history. The law assumes that all the strain changes resulting from stress changes are instantaneous and the system is completely reversible and all the input energy is recovered in unloading.

In case of uniaxial loading, stress is related to strain as

$$\sigma_x = E\varepsilon_x \tag{4.1}$$

where E is known as "Young's modulus" or "Modulus of Elasticity".

Expression (4.1) is applicable within the linear elastic range and is called Hooke's law.

In general, each strain is dependent on each stress. For example, the strain ε_x written as a function of each stress is

$$\varepsilon_x = C_{11}\sigma_x + C_{12}\sigma_y + C_{13}\sigma_z + C_{14}\tau_{xy} + C_{15}\tau_{yz}$$
$$+ C_{16}\tau_{xx} + C_{17}\tau_{xz} + C_{18}\tau_{zy} + C_{19}\tau_{yx} \tag{4.2}$$

Similarly, stresses can be expressed in terms of strains stating that at each point in a material, each stress component is linearly related to all the strain components. This is known as "Generalized Hooke's Law".

Hence,

$$\sigma_x = D_{11}\varepsilon_x + D_{12}\varepsilon_y + D_{13}\varepsilon_z + D_{14}\gamma_{xy} + D_{15}\gamma_{yz}$$
$$+ D_{16}\gamma_{zx} + D_{17}\gamma_{xz} + D_{18}\gamma_{zy} + D_{19}\gamma_{yx} \tag{4.3}$$

For the most general case of three-dimensional state of stress, Eq. (4.3) can be written as

$$\{\sigma_{ij}\} = (D_{ijkl})\{\varepsilon_{kl}\} \tag{4.4}$$

where

(D_{ijkl}) Elasticity matrix
$\{\sigma_{ij}\}$ Stress components
$\{\varepsilon_{kl}\}$ Strain components

Since both stress σ_{ij} and strain ε_{ij} are second-order tensors, it follows that D_{ijkl} is a fourth-order tensor, which consists of $3^4 = 81$ material constants if symmetry is not assumed. Therefore in matrix notation, the stress–strain relations would be

$$
\begin{Bmatrix}
\sigma_x \\
\sigma_y \\
\sigma_z \\
\tau_{xy} \\
\tau_{yz} \\
\tau_{zx} \\
\tau_{xz} \\
\tau_{zy} \\
\tau_{yx}
\end{Bmatrix}
=
\begin{bmatrix}
D_{11} & D_{12} & D_{13} & D_{14} & D_{15} & D_{16} & D_{17} & D_{18} & D_{19} \\
D_{21} & D_{22} & D_{23} & D_{24} & D_{25} & D_{26} & D_{27} & D_{28} & D_{29} \\
D_{31} & D_{32} & D_{33} & D_{34} & D_{35} & D_{36} & D_{37} & D_{38} & D_{39} \\
D_{41} & D_{42} & D_{43} & D_{44} & D_{45} & D_{46} & D_{47} & D_{48} & D_{49} \\
D_{51} & D_{52} & D_{53} & D_{54} & D_{55} & D_{56} & D_{57} & D_{58} & D_{59} \\
D_{61} & D_{62} & D_{63} & D_{64} & D_{65} & D_{66} & D_{67} & D_{68} & D_{69} \\
D_{71} & D_{72} & D_{73} & D_{74} & D_{75} & D_{76} & D_{77} & D_{78} & D_{79} \\
D_{81} & D_{82} & D_{83} & D_{84} & D_{85} & D_{86} & D_{87} & D_{88} & D_{89} \\
D_{91} & D_{92} & D_{93} & D_{94} & D_{95} & D_{96} & D_{97} & D_{98} & D_{99}
\end{bmatrix}
\begin{Bmatrix}
\varepsilon_x \\
\varepsilon_y \\
\varepsilon_z \\
\gamma_{xy} \\
\gamma_{yz} \\
\gamma_{zx} \\
\gamma_{xz} \\
\gamma_{zy} \\
\gamma_{yx}
\end{Bmatrix}
\tag{4.5}
$$

Now, from $\sigma_{ij} = \sigma_{ji}$ and $\varepsilon_{ij} = \varepsilon_{ji}$ (i.e. shear is normally symmetric) the number of 81 material constants is reduced to 36 under symmetric conditions of $D_{ijkl} = D_{jikl} = D_{ijlk} = D_{jilk}$

Therefore in matrix notation, the stress–strain relations can be

$$
\begin{Bmatrix}
\sigma_x \\
\sigma_y \\
\sigma_z \\
\tau_{xy} \\
\tau_{yz} \\
\tau_{zx}
\end{Bmatrix}
=
\begin{bmatrix}
D_{11} & D_{12} & D_{13} & D_{14} & D_{15} & D_{16} \\
D_{21} & D_{22} & D_{23} & D_{24} & D_{25} & D_{26} \\
D_{31} & D_{32} & D_{33} & D_{34} & D_{35} & D_{36} \\
D_{41} & D_{42} & D_{43} & D_{44} & D_{45} & D_{46} \\
D_{51} & D_{52} & D_{53} & D_{54} & D_{55} & D_{56} \\
D_{61} & D_{62} & D_{63} & D_{64} & D_{65} & D_{66}
\end{bmatrix}
\begin{Bmatrix}
\varepsilon_x \\
\varepsilon_y \\
\varepsilon_z \\
\gamma_{xy} \\
\gamma_{yz} \\
\gamma_{zx}
\end{Bmatrix}
\tag{4.6}
$$

Equation (4.6) indicates that 36 elastic constants are necessary for the most general form of anisotropy (different elastic properties in all directions). It is generally accepted, however, that the stiffness matrix D_{ij} is symmetric, in which case the number of independent elastic constants will be reduced to 21. This can be shown by assuming the existence of a strain energy function U.

It is often desired in classical elasticity to have a potential function

$$
U = U\left(\varepsilon_{ij}\right) \tag{4.7}
$$

With the property that

$$
\frac{\partial U}{\partial \varepsilon_{ij}} = \sigma_{ij} \tag{4.8}
$$

Such a function is called a "strain energy" or "strain energy density function". By Eq. (4.8), we can write

$$
\frac{\partial U}{\partial \varepsilon_i} = \sigma_i = D_{ij}\,\varepsilon_j \tag{4.9}
$$

Differentiating Eq. (4.9) with respect to ε_j, then

$$\frac{\partial^2 U}{\partial \varepsilon_i \partial \varepsilon_j} = D_{ij} \tag{4.10}$$

The free index in Eq. (4.9) can be changed so that

$$\frac{\partial U}{\partial \varepsilon_j} = \sigma_j = D_{ji} \varepsilon_i \tag{4.11}$$

Differentiating Eq. (4.11) with respect to ε_i, then,

$$\frac{\partial^2 U}{\partial \varepsilon_j \varepsilon_i} = D_{ji} \tag{4.12}$$

Hence, Eqs. (4.10) and (4.12) are equal, or $D_{ij} = D_{ji}$
which implies that D_{ij} is symmetric. Then most general form of the stiffness matrix or array becomes

$$\begin{Bmatrix} \sigma_x \\ \sigma_y \\ \sigma_z \\ \tau_{xy} \\ \tau_{yz} \\ \tau_{zx} \end{Bmatrix} = \begin{bmatrix} D_{11} & D_{12} & D_{13} & D_{14} & D_{15} & D_{16} \\ D_{12} & D_{22} & D_{23} & D_{24} & D_{25} & D_{26} \\ D_{13} & D_{23} & D_{33} & D_{34} & D_{35} & D_{36} \\ D_{14} & D_{24} & D_{34} & D_{44} & D_{45} & D_{46} \\ D_{15} & D_{25} & D_{35} & D_{45} & D_{55} & D_{56} \\ D_{16} & D_{26} & D_{36} & D_{46} & D_{56} & D_{66} \end{bmatrix} \begin{Bmatrix} \varepsilon_x \\ \varepsilon_y \\ \varepsilon_z \\ \gamma_{xy} \\ \gamma_{yz} \\ \gamma_{zx} \end{Bmatrix} \tag{4.13}$$

Or

$$\begin{Bmatrix} \sigma_x \\ \sigma_y \\ \sigma_z \\ \tau_{xy} \\ \tau_{yz} \\ \tau_{zx} \end{Bmatrix} = \begin{bmatrix} D_{11} & D_{12} & D_{13} & D_{14} & D_{15} & D_{16} \\ & D_{22} & D_{23} & D_{24} & D_{25} & D_{26} \\ & & D_{33} & D_{34} & D_{35} & D_{36} \\ & & & D_{44} & D_{45} & D_{46} \\ & & & & D_{55} & D_{56} \\ & & & & & D_{66} \end{bmatrix} \begin{Bmatrix} \varepsilon_x \\ \varepsilon_y \\ \varepsilon_z \\ \gamma_{xy} \\ \gamma_{yz} \\ \gamma_{zx} \end{Bmatrix} \tag{4.14}$$

Further, a material that exhibits symmetry with respect to three mutually orthogonal planes is called an "orthotropic" material. If the xy, yz and zx planes are considered planes of symmetry, then Eq. (4.13) reduces to 12 elastic constants as below.

$$\begin{Bmatrix} \sigma_x \\ \sigma_y \\ \sigma_z \\ \tau_{xy} \\ \tau_{yz} \\ \tau_{zx} \end{Bmatrix} = \begin{bmatrix} D_{11} & D_{12} & D_{13} & 0 & 0 & 0 \\ D_{21} & D_{22} & D_{23} & 0 & 0 & 0 \\ D_{31} & D_{32} & D_{33} & 0 & 0 & 0 \\ 0 & 0 & 0 & D_{44} & 0 & 0 \\ 0 & 0 & 0 & 0 & D_{55} & 0 \\ 0 & 0 & 0 & 0 & 0 & D_{66} \end{bmatrix} \begin{Bmatrix} \varepsilon_x \\ \varepsilon_y \\ \varepsilon_z \\ \gamma_{xy} \\ \gamma_{yz} \\ \gamma_{zx} \end{Bmatrix} \tag{4.15}$$

Also, due to orthotropic symmetry, the number of material constants for a linear elastic orthotropic material reduces to 9 as shown below.

$$
\begin{Bmatrix} \sigma_x \\ \sigma_y \\ \sigma_z \\ \tau_{xy} \\ \tau_{yz} \\ \tau_{zx} \end{Bmatrix} = \begin{bmatrix} D_{11} & D_{12} & D_{13} & 0 & 0 & 0 \\ & D_{22} & D_{23} & 0 & 0 & 0 \\ & & D_{33} & 0 & 0 & 0 \\ & & & D_{44} & 0 & 0 \\ & & & & D_{55} & 0 \\ & & & & & D_{66} \end{bmatrix} \begin{Bmatrix} \varepsilon_x \\ \varepsilon_y \\ \varepsilon_z \\ \gamma_{xy} \\ \gamma_{yz} \\ \gamma_{zx} \end{Bmatrix} \qquad (4.16)
$$

Now, in the case of a transversely isotropic material, the material exhibits a rationally elastic symmetry about one of the co-ordinate axes, x, y and z. In such case, the material constants reduce to 8 as shown below.

$$
\begin{Bmatrix} \sigma_x \\ \sigma_y \\ \sigma_z \\ \tau_{xy} \\ \tau_{yz} \\ \tau_{zx} \end{Bmatrix} = \begin{bmatrix} D_{11} & D_{12} & D_{13} & 0 & 0 & 0 \\ & D_{22} & D_{23} & 0 & 0 & 0 \\ & & D_{33} & 0 & 0 & 0 \\ & & & \frac{1}{2}(D_{11} - D_{12}) & 0 & 0 \\ \text{Symmetry} & & & & D_{55} & 0 \\ & & & & & D_{66} \end{bmatrix} \begin{Bmatrix} \varepsilon_x \\ \varepsilon_y \\ \varepsilon_z \\ \gamma_{xy} \\ \gamma_{yz} \\ \gamma_{zx} \end{Bmatrix} \qquad (4.17)
$$

Further, for a linearly elastic material with cubic symmetry for which the properties along the x-, y- and z-directions are identical, there are only 3 independent material constants. Therefore, the matrix form of the stress–strain relation can be expressed as:

$$
\begin{Bmatrix} \sigma_x \\ \sigma_y \\ \sigma_z \\ \tau_{xy} \\ \tau_{yz} \\ \tau_{zx} \end{Bmatrix} = \begin{bmatrix} D_{11} & D_{12} & D_{12} & 0 & 0 & 0 \\ & D_{11} & D_{12} & 0 & 0 & 0 \\ & & D_{11} & 0 & 0 & 0 \\ & & & D_{44} & 0 & 0 \\ \text{Symmetry} & & & & D_{44} & 0 \\ & & & & & D_{44} \end{bmatrix} \begin{Bmatrix} \varepsilon_x \\ \varepsilon_y \\ \varepsilon_z \\ \gamma_{xy} \\ \gamma_{yz} \\ \gamma_{zx} \end{Bmatrix} \qquad (4.18)
$$

Isotropy

For a material whose elastic properties are not a function of direction at all, only two independent elastic material constants are sufficient to describe its behaviour completely. This material is called "isotropic linear elastic". The stress–strain relationship for this material is thus written as an extension of that of a transversely isotropic material as shown below.

$$
\begin{Bmatrix} \sigma_x \\ \sigma_y \\ \sigma_z \\ \tau_{xy} \\ \tau_{yz} \\ \tau_{zx} \end{Bmatrix}
=
\begin{bmatrix}
D_{11} & D_{12} & D_{12} & 0 & 0 & 0 \\
 & D_{11} & D_{12} & 0 & 0 & 0 \\
 & & D_{11} & 0 & 0 & 0 \\
 & & & \frac{1}{2}(D_{11}-D_{12}) & 0 & 0 \\
 & & & & \frac{1}{2}(D_{11}-D_{12}) & 0 \\
 & & & & & \frac{1}{2}(D_{11}-D_{12})
\end{bmatrix}
\begin{Bmatrix} \varepsilon_x \\ \varepsilon_y \\ \varepsilon_z \\ \gamma_{xy} \\ \gamma_{yz} \\ \gamma_{zx} \end{Bmatrix}
$$

$$(4.19)$$

Thus, we get only 2 independent elastic constants.

Replacing D_{12} and $\frac{1}{2}(D_{11}-D_{12})$, respectively, by λ and G which are called "Lame's constants", where G is also called shear modulus of elasticity, Eq. (4.19) can be written as:

$$
\begin{Bmatrix} \sigma_x \\ \sigma_y \\ \sigma_z \\ \tau_{xy} \\ \tau_{yz} \\ \tau_{zx} \end{Bmatrix}
=
\begin{bmatrix}
2G+\lambda & \lambda & \lambda & 0 & 0 & 0 \\
 & 2G+\lambda & \lambda & 0 & 0 & 0 \\
 & & 2G+\lambda & 0 & 0 & 0 \\
 & & & G & 0 & 0 \\
 & \text{Symmetry} & & & G & 0 \\
 & & & & & G
\end{bmatrix}
\begin{Bmatrix} \varepsilon_x \\ \varepsilon_y \\ \varepsilon_z \\ \gamma_{xy} \\ \gamma_{yz} \\ \gamma_{zx} \end{Bmatrix}
$$

$$(4.20)$$

Therefore, the stress–strain relationships may be expressed as

$$
\begin{Bmatrix} \sigma_x \\ \sigma_y \\ \sigma_z \\ \tau_{xy} \\ \tau_{yz} \\ \tau_{zx} \end{Bmatrix}
=
\begin{bmatrix}
2G+\lambda & \lambda & \lambda & 0 & 0 & 0 \\
\lambda & 2G+\lambda & \lambda & 0 & 0 & 0 \\
\lambda & \lambda & 2G+\lambda & 0 & 0 & 0 \\
0 & 0 & 0 & G & 0 & 0 \\
0 & 0 & 0 & 0 & G & 0 \\
0 & 0 & 0 & 0 & 0 & G
\end{bmatrix}
\begin{Bmatrix} \varepsilon_x \\ \varepsilon_y \\ \varepsilon_z \\ \gamma_{xy} \\ \gamma_{yz} \\ \gamma_{zx} \end{Bmatrix}
$$

$$(4.21)$$

Therefore,

$$
\begin{aligned}
\sigma_x &= (2G+\lambda)\varepsilon_x + \lambda\left(\varepsilon_y + \varepsilon_z\right) \\
\sigma_y &= (2G+\lambda)\varepsilon_y + \lambda(\varepsilon_z + \varepsilon_x) \\
\sigma_z &= (2G+\lambda)\varepsilon_z + \lambda\left(\varepsilon_x + \varepsilon_y\right)
\end{aligned}
$$

$$(4.22)$$

Also,

$$
\begin{aligned}
\tau_{xy} &= G\gamma_{xy} \\
\tau_{yz} &= G\gamma_{yz} \\
\tau_{zx} &= G\gamma_{zx}
\end{aligned}
$$

Now, expressing strains in terms of stresses, we get

$$\varepsilon_x = \frac{\lambda + G}{G(3\lambda + 2G)}\sigma_x - \frac{\lambda}{2G(3\lambda + 2G)}(\sigma_y + \sigma_z)$$

$$\varepsilon_y = \frac{\lambda + G}{G(3\lambda + 2G)}\sigma_y - \frac{\lambda}{2G(3\lambda + 2G)}(\sigma_z + \sigma_x)$$

$$\varepsilon_z = \frac{\lambda + G}{G(3\lambda + 2G)}\sigma_z - \frac{\lambda}{2G(3\lambda + 2G)}(\sigma_x + \sigma_y) \tag{4.23}$$

$$\gamma_{xy} = \frac{\tau_{xy}}{G}$$

$$\gamma_{yz} = \frac{\tau_{yz}}{G}$$

$$\gamma_{zx} = \frac{\tau_{zx}}{G}$$

Now consider a simple tensile test
Therefore,

$$\varepsilon_x = \frac{\sigma_x}{E} = \frac{\lambda + G}{G(3\lambda + 2G)} = \sigma_x$$

$$\text{or} \quad \frac{1}{E} = \frac{\lambda + G}{G(3\lambda + 2G)}$$

$$\text{or} \quad E = \frac{G(3\lambda + 2G)}{(\lambda + G)} \tag{4.24}$$

where E = Modulus of Elasticity
    Also,

$$\varepsilon_y = -\nu\varepsilon_x = -\nu\frac{\sigma_x}{E}$$

$$\varepsilon_{\bar{z}} = -\nu\varepsilon_x = -\nu\frac{\sigma_x}{E}$$

where ν = Poisson's ratio
    For $\sigma_y = \sigma_z = 0$, we get from Eq. (4.23)

$$-\frac{\lambda}{2G(3\lambda + 2G)}\sigma_x = -\frac{\nu}{E}\sigma_x$$

Therefore,

$$\frac{\nu}{E} = \frac{\lambda}{2G(3\lambda + 2G)} \tag{4.25}$$

Substituting the value of E from Eq. (4.24), we get

$$\frac{v(\lambda + G)}{G(3\lambda + 2G)} = \frac{\lambda}{2G(3\lambda + 2G)}$$

Therefore,

$$2v(\lambda + G) = \lambda$$

$$\text{or} \quad v = \frac{\lambda}{2(\lambda + G)} \qquad (4.26)$$

Solving for λ from Eqs. (4.23) and (4.24), we get

$$\lambda = \frac{G(2G - E)}{(E - 3G)} = \frac{4G^2 v}{(E - 6Gv)}$$

$$\text{or} \quad G = \frac{E}{2(1 + v)} \qquad (4.27)$$

For a hydrostatic stress, i.e. all round compression p,

$$\sigma_x = \sigma_y = \sigma_z = -p$$

Therefore,

$$\varepsilon_x + \varepsilon_y + \varepsilon_z = \frac{-3(1 - 2v)p}{E}$$

$$\text{or} \quad -p = \frac{E(\varepsilon_x + \varepsilon_y + \varepsilon_z)}{3(1 - 2v)}$$

$$= (\lambda + \frac{2G}{3})(\varepsilon_x + \varepsilon_y + \varepsilon_z)$$

$$\text{or} \quad -p = K(\varepsilon_x + \varepsilon_y + \varepsilon_z)$$

Hence,

$$K = \left(\lambda + \frac{2G}{3}\right) \qquad (4.28)$$

where $K = $ Bulk modulus of elasticity.

Also,

$$-p = K(\varepsilon_x + \varepsilon_y + \varepsilon_z)$$

$$-p = K\left[\frac{-3p(1-2v)}{E}\right]$$

or $E = K[3(1-2v)]$

Therefore,

$$K = \frac{E}{3(1-2v)} \tag{4.29}$$

Lames constant λ can be denoted as

$$\lambda = \frac{v\,E}{(1+v)\,(1-2v)} \tag{4.30}$$

Thus among the four elastic constants E, v, G and K only two constants can be replaced to the other two ones. Therefore, any two of these constants can be taken as the independent constants. Further, it is to be noted from Eq. (4.29) that for the bulk modulus to be positive, the value of Poisson's ratio v cannot be greater than 0.5. Hence, for $v = 0.5$, Eq. (4.29) becomes

$$K = \infty \quad \text{and} \quad G = \frac{E}{3} \tag{4.31}$$

Hence, materials having Poisson's ratio equal to 0.5 are called as incompressible materials, since, for such materials, the volumetric stain is zero. Table 4.1 gives the relations between various elastic constants and Table 4.2 presents typical values of elastic constants of some materials.

4.3 Elastic Strain Energy for Uniaxial Stress

In mechanics, energy is defined as the capacity to do work, and work is the product of force and the distance, in the direction, the force moves. In solid, deformable bodies, stresses multiplied by their respective areas, forces and deformations are distances. The product of these two quantities is the internal work done in a body by externally applied forces. This internal work is stored in a body as the internal elastic energy of deformation or the elastic strain energy.

Consider an infinitesimal element as shown in Fig. 4.1a, subjected to a normal stress σ_x. The force acting on the right or the left face of this element is $\sigma_x dy dz$. This force causes an elongation in the element by an amount $\varepsilon_x dx$, where ε_x is the

Table 4.1 Relation between elastic constants

	E	v	K	G	λ
$E.v$	E	v	$\frac{E}{3(1+2v)}$	$\frac{E}{2(1+v)}$	$\frac{Ev}{(1+v)(1-2v)}$
$E.K$	E	$\frac{3K-E}{6K}$	K	$\frac{3KE}{9K-E}$	$\frac{3K(3K-E)}{9K-E}$
$E.G$	E	$\frac{E-2G}{2G}$	$\frac{GE}{3(3G-E)}$	G	$\frac{G(E-2G)}{3G-E}$
$E.\lambda$	E	$\frac{R-E-\lambda}{4\lambda}$	$\frac{R+E+3\lambda}{6}$	$\frac{R+E-3\lambda}{4}$	λ
$v.K$	$3k(1-2v)$	v	K	$\frac{3K(1-2v)}{2(1+v)}$	$\frac{3Kv}{1+v}$
$v.G$	$2G(1+v)$	v	$\frac{2G(1+v)}{3(1-2v)}$	G	$\frac{2Gv}{1-2v}$
$v.\lambda$	$\frac{\lambda(1+v)(1-2v)}{v}$	v	$\frac{\lambda(1+v)}{3v}$	$\frac{\lambda(1-2v)}{2v}$	λ
$K.G$	$\frac{9KG}{3K+G}$	$\frac{3K-2G}{6K+2G}$	K	G	$K-\frac{2}{3}G$
$K.\lambda$	$\frac{9KG(k-\lambda)}{3K-\lambda}$	$\frac{\lambda}{3K-\lambda}$	K	$\frac{3K-\lambda}{2}$	λ
$G.\lambda$	$\frac{G(3\lambda+2G)}{\lambda+G}$	$\frac{\lambda}{2(\lambda+G)}$	$\frac{3\lambda+2G}{3}$	G	λ

$$R = \sqrt{E^2 + 9\lambda^2 + 2E\lambda} > 0$$

Table 4.2 Typical values of elastic moduli for common engineering materials

	E(GPa)	v	G(GPa)	λ(GPa)	k(GPa)	$\alpha(10^{-6/\circ}C)$
Aluminium	68.9	0.34	25.7	54.6	71.8	25.5
Concrete	27.6	0.20	11.5	7.7	15.3	11
Copper	89.6	0.34	33.4	71	93.3	18
Glass	68.9	0.25	27.6	27.6	45.9	8.8
Nylon	28.3	0.40	10.1	4.04	47.2	102
Rubber	0.0019	0.499	0.654×10^{-3}	0.326	0.326	200
Steel	207	0.29	80.2	111	164	13.5

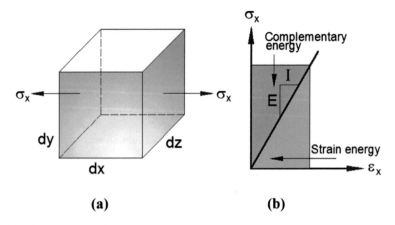

(a)   **(b)**

Fig. 4.1 Element subjected to a normal stress

strain in the direction x. The average force acting on the element while deformation is taking place is $\sigma_x \frac{dy\,dz}{2}$. This average force multiplied by the distance through which it acts is the work done on the element. For a perfectly elastic body no energy is dissipated, and the work done on the element is stored as recoverable internal strain energy. Therefore, the internal elastic strain energy U for an infinitesimal element subjected to uniaxial stress is

$$dU = \frac{1}{2}\sigma_x dy\,dz \times \varepsilon_x dx$$
$$= \frac{1}{2}\sigma_x \varepsilon_x dx\,dy\,dz$$

Therefore,

$$dU = \frac{1}{2}\sigma_x \varepsilon_x dV \tag{4.32}$$

where

$dV = dx\,dy\,dz$ volume of the element.

Thus, the above expression gives the strain energy stored in an elastic body per unit volume of the material, which is called strain energy density U_0.

Hence,

$$\frac{dU}{dV} = U_0 = \frac{1}{2}\sigma_x \varepsilon_x \tag{4.33}$$

The above expression may be graphically interpreted as an area under the inclined line on the stress–strain diagram as shown in Fig. 4.1b. The area enclosed by the inclined line and the vertical axis is called the complementary energy. For linearly elastic materials, the two areas are equal.

4.4   Strain Energy in an Elastic Body

When work is done by an external force on certain systems, their internal geometric states are altered in such a way that they have the potential to give back equal amounts of work whenever they are returned to their original configurations. Such systems are called conservative, and the work done on them is said to be stored in the form of potential energy. For example, the work done in lifting a weight is said to be stored as a gravitational potential energy. The work done in deforming an elastic spring is said to be stored as elastic potential energy. By contrast, the work done in sliding a block against friction is not recoverable; i.e. friction is a non-conservative mechanism.

Now we can extend the concept of elastic strain energy to arbitrary linearly elastic bodies subjected to small deformations.

Figure 4.2a shows a uniaxial stress component σ_x acting on a rectangular element, and Fig. 4.2b shows the corresponding deformation including the elongation due to the strain component ε_x. The elastic energy stored in such an element is commonly called strain energy.

In this case, the force $\sigma_x dy dz$ acting on the positive x-face does work as the element undergoes the elongation $\varepsilon_x dx$. In a linearly elastic material, strain grows in proportion to stress. Thus, the strain energy dU stored in the element, when the final values of stress and strain are σ_x and ε_x is

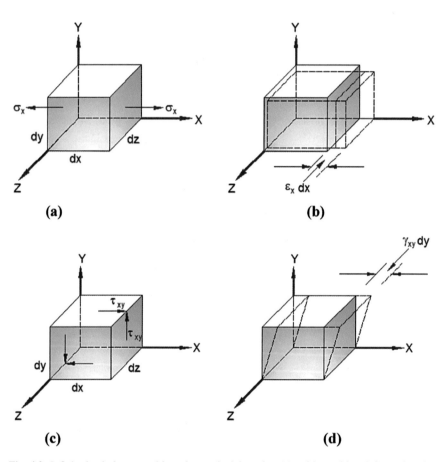

Fig. 4.2 Infinitesimal element subjected to: uniaxial tension (**a**), with resulting deformation (**b**); pure shear (**c**), with resulting deformation (**d**)

$$dU = \frac{1}{2}(\sigma_x dy dz)(\varepsilon_x dx)$$

$$= \frac{1}{2}\sigma_x \varepsilon_x dV \tag{4.34}$$

where

$dV = dx\ dy\ dz$ volume of the element.

If an elastic body of total volume V is made up of such elements, the total strain energy U is obtained by integration

$$U = \frac{1}{2}\int_v \sigma_x \varepsilon_y dV \tag{4.35}$$

Taking $\sigma_x = \frac{P}{A}$ and $\varepsilon_x = \frac{\delta}{L}$
where

P uniaxial load on the member
δ displacement due to load P
L length of the member,
A cross-sectional area of the member

We can write Eq. (4.28) as

$$U = \frac{1}{2}\left(\frac{P}{A}\right)\left(\frac{\delta}{L}\right)\int_v dV$$

Therefore,

$$U = \frac{1}{2}P.\delta \quad \text{since } V = L \times A \tag{4.36}$$

Next consider the shear stress component τ_{xy} acting on an infinitesimal element in Fig. 4.2c. The corresponding deformation due to the shear strain component γ_{xy} is indicated in Fig. 4.2d. In this case, the force $\tau_{xy}dxdz$ acting on the positive y face does work as that face translates through the distance $\gamma_{xy}\ dy$. Because of linearity, γ_{xy} and τ_{xy} grow in proportion as the element is deformed.

The strain energy stored in the element, when the final values of strain and stress are γ_{xy} and τ_{xy} is

$$dU = \frac{1}{2}(\tau_{xy}dxdz)(\gamma_{xy}dy)$$

$$= \frac{1}{2}\tau_{xy}\gamma_{xy}dxdydz$$

Therefore,

$$dU = \frac{1}{2}\tau_{xy}\gamma_{xy}dV \tag{4.37}$$

The results are analogous to Eqs. (4.34), and (4.37) can be written for any other pair of stress and strain components (e.g. σ_y and ε_y or τ_{yz} and γ_{yz}) whenever the stress component involved is the only stress acting on the element.

Finally, we consider a general state of stress in which all six stress components are present. The corresponding deformation will in general involve all six strain components. The total strain energy stored in the element when the final stresses are $\sigma_x, \sigma_y, \sigma_z, \tau_{xy}, \tau_{yz}, \tau_{zx}$ and the final strains are $\varepsilon_x, \varepsilon_y, \varepsilon_z, \gamma_{xy}, \gamma_{yz}, \gamma_{zx}$ is thus

$$dU = \frac{1}{2}(\sigma_x\varepsilon_x + \sigma_y\varepsilon_y + \sigma_z\varepsilon_z + \tau_{xy}\gamma_{xy} + \tau_{yz}\gamma_{yz} + \tau_{zx}\gamma_{zx})dV \tag{4.38}$$

In general, the final stresses and strains vary from point to point in the body. The strain energy stored in the entire body is obtained by integrating Eq. (4.32) over the volume of the body.

$$U = \frac{1}{2}\int_v (\sigma_x\varepsilon_x + \sigma_y\varepsilon_y + \sigma_z\varepsilon_z + \tau_{xy}\gamma_{xy} + \tau_{yz}\gamma_{yz} + \tau_{zx}\gamma_{zx})dV \tag{4.39}$$

The above formula for strain energy applies to small deformations of any linearly elastic body.

4.5   Boundary Conditions

The boundary conditions are specified in terms of surface forces on certain boundaries of a body to solve problems in continuum mechanics. When the stress components vary over the volume of the body, they must be in equilibrium with the externally applied forces on the boundary of the body. Thus, the external forces may be regarded as a continuation of internal stress distribution.

Consider a two-dimensional body as shown in Fig. 4.3

Take a small triangular prism ABC, so that the side BC coincides with the boundary of the plate. At a point P on the boundary, the outward normal is n. Let $\overline{X}$ and $\overline{Y}$ be the components of the surface forces per unit area at this point of boundary. $\overline{X}$ and $\overline{Y}$ must be a continuation of the stresses σ_x, σ_y and τ_{xy} at the boundary. Now, using Cauchy's equation, we have

$$T_x = \overline{X} = \sigma_x l + \tau_{xy}m$$
$$T_y = \overline{Y} = \tau_{xy}l + \sigma_y m \tag{a}$$

in which l and m are the direction cosines of the normal n to the boundary.

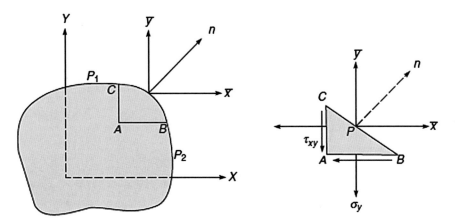

Fig. 4.3 An element at the boundary of a body

For a particular case of a rectangular plate, the co-ordinate axes are usually taken parallel to the sides of the plate and the boundary conditions (equation a) can be simplified. For example, if the boundary of the plate is parallel to x-axis, at point P_1, then the boundary conditions become

$$\overline{X} = \tau_{xy} \text{ and } \overline{Y} = \sigma_y \tag{b}$$

Further, if the boundary of the plate is parallel to y-axis, at point P_2, then the boundary conditions become

$$\overline{X} = \sigma_x \text{ and } \overline{Y} = \tau_{xy} \tag{c}$$

It is seen from the above that at the boundary, the stress components become equal to the components of surface forces per unit area of the boundary.

Similarly, the boundary conditions in three-dimensional problems can be considered as follows:

$$\overline{X} = \sigma_x l + \tau_{xy} m + \tau_{xz} n$$
$$\overline{Y} = \tau_{xy} l + \sigma_y m + \tau_{yz} n \tag{d}$$
$$\overline{Z} = \tau_{xz} l + \tau_{y_z} m + \sigma_z n$$

4.6 ST. Venant's Principle

For the purpose of analysing the statics or dynamics of a body, one force system may be replaced by an equivalent force system whose force and moment resultants

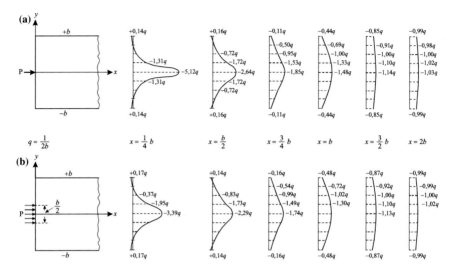

Fig. 4.4 Surface of a body subjected to **a** concentrated load and **b** strip load of width b

are identical. Such force resultants, while equivalent need not cause an identical distribution of strain, owing to difference in the arrangement of forces. St. Venant's principle permits the use of an equivalent loading for the calculation of stress and strain.

St. Venant's principle states that if a certain system of forces acting on a portion of the surface of a body is replaced by a different system of forces acting on the same portion of the body, then the effects of the two different systems at locations sufficiently far distant from the region of application of forces, are essentially the same, provided that the two systems of forces are statically equivalent (i.e. the same resultant force and the same resultant moment).

St. Venant's principle is very convenient and useful in obtaining solutions to many engineering problems in elasticity. The principle helps to the great extent in prescribing the boundary conditions very precisely when it is very difficult to do so. The following Figs. 4.4, 4.5 and 4.6 illustrate the St. Venant's principle.

Figures 4.4, 4.5 and 4.6 demonstrate the distribution of stresses (q) in the body when subjected to various types of loading. In all the cases, the distribution of stress throughout the body is altered only near the regions of load application. However, the stress distribution is not altered at a distance $x = 2b$ irrespective of loading conditions.

4.7   Principle of Superposition

It is to be noted that all the governing equations of elasticity developed in the previous sections are valid for small deformations and hence they are linear. Therefore, any

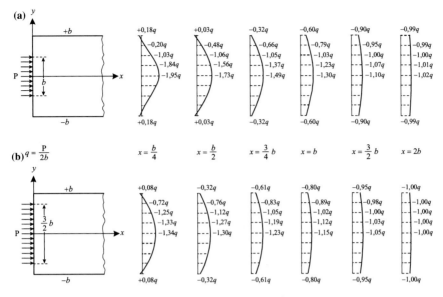

Fig. 4.5 Surface of a body subjected to **a** strip load of width b and **b** strip load of width $1.5b$

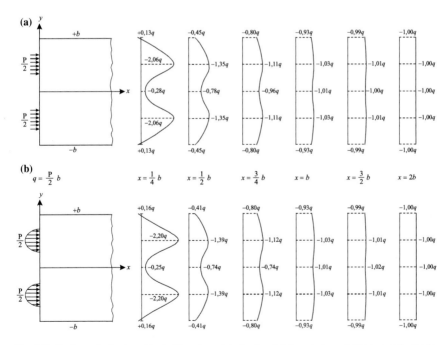

Fig. 4.6 Surface of a body subjected to **a** two strip load and **b** inverted parabolic two strip loads

linear combinations of functions that individually satisfy the equations of elasticity
also satisfy the equations in combined form. This is generally called the principle of
superposition.

Let $\sigma_x, \sigma_y, \ldots \tau_{xy}, \tau_{yz}, \ldots$ be the components of stress so determined in a body
due to a set of surface forces $\bar{X}, \bar{Y}, \bar{Z}$ and body forces F_x, F_y, F_z.

Let $\sigma_x', \sigma_y', \ldots \tau_{xy}', \tau_{yz}', \ldots$ be the components of stress in the same body due to a
different set of surface forces $\bar{X}', \bar{Y}', \bar{Z}'$ and body forces F_x', F_y', F_z'.

Now the stress components $\sigma_x + \sigma_x', \sigma_y + \sigma_y' \ldots \tau_{xy} + \tau_{xy}' \ldots$ will represent the
stresses due to surface forces $\bar{X} + \bar{X}', \ldots$ and the body forces $F_x + F_x' \ldots$

Now, consider the Eq. (2.32a) and rewriting we get:

$$\frac{\partial \sigma_x'}{\partial x} + \frac{\partial \tau_{xy}'}{\partial y} + \frac{\partial \tau_{xz}'}{\partial z} + F_x' = 0$$

Adding to the corresponding equation, then

$$\frac{\partial}{\partial x}\left(\sigma_x + \sigma_x'\right) + \frac{\partial}{\partial y}\left(\tau_{xy} + \tau_{xy}'\right) + \frac{\partial}{\partial z}\left(\tau_{xz} + \tau_{xy}'\right) + F_x + F_x' = 0 \qquad (4.40)$$

Therefore, it can be stated that adding up the two sets of equations of equilibrium
will provide us another set of equations of equilibrium corresponding to the combined
state of stress. The same procedure can be employed to compatibility conditions also.

It should be realised that the principle of superposition will not hold good large
deformations. There are several cases even for small deformations under certain
conditions the principle of superposition is not valid.

4.8 Existence and Uniqueness of Solution (Uniqueness Theorem)

This is an important theorem in the theory of elasticity and distinguishes elastic deformations from plastic deformations. The theorem states that, for every problem of elasticity defined by a set of governing equations and boundary conditions, there exists
one and only one solution. This means that "elastic problems have a unique solution"
and two different solutions cannot satisfy the same set of governing equations and
boundary conditions.

Proof In proving the above theorem, one must remember that only elastic problems
are dealt with infinitesimal strains and displacements. If the strains and displacements
are not infinitesimal, the solution may not be unique.

Let a set of stresses $\sigma_x', \sigma_y', \ldots\ldots\ldots\tau_{zx}'$ represents a solution for the equilibrium of
a body under surface forces X, Y, Z and body forces F_x, F_y, F_z. Then the equations
of equilibrium and boundary conditions must be satisfied by these stresses, giving

$$\frac{\partial \sigma_x'}{\partial x} + \frac{\partial \tau_{xy}'}{\partial y} + \frac{\partial \tau_{xz}'}{\partial z} + x = 0; (x, y, z)$$

and

$$\sigma_x' l + \tau_{xy}' m + \tau_{xz}' n = F_x; (x, y, z)$$

where (x, y, z) means that there are two more equations obtained by changing the suffixes y for x and z for y, in a cyclic order.

Similarly, if there is another set of stresses $\sigma_x'', \sigma_y'', \ldots \tau_{zx}''$ which also satisfies the boundary conditions and governing equations we have,

$$\frac{\partial \sigma_x''}{\partial x} + \frac{\partial \tau_{xy}''}{\partial y} + \frac{\partial \tau_{xz}''}{\partial z} + x = 0; (x, y, z)$$

and

$$\sigma_x'' l + \tau_{xy}'' m + \tau_{xz}'' n = F_x; (x, y, z)$$

By subtracting the equations of the above set from the corresponding equations of the previous set, we get the following set,

$$\frac{\partial}{\partial x}(\sigma_x' - \sigma_x'') + \frac{\partial}{\partial y}(\tau_{xy}' - \tau_{xy}'') + \frac{\partial}{\partial z}(\tau_{xz}' - \tau_{xz}'') = 0; (x, y, z)$$

and

$$(\sigma_x' - \sigma_x'') l + (\tau_{xy}' - \tau_{xy}'') m + (\tau_{xz}' - \tau_{xz}'') n = 0; (x, y, z)$$

In the same way, it is shown that the new strain components $(\varepsilon_x' - \varepsilon_x''), (\varepsilon_y' - \varepsilon_y'') \ldots,$ etc. also satisfy the equations of compatibility. A new solution $(\sigma_x' - \sigma_x''), (\sigma_x' - \sigma_y''), \ldots (\tau_{xz}' - \tau_{xz}'')$ represents a situation where body forces and surface forces both are zero. The work done by these forces during loading is zero, and it follows that the total strain energy vanishes, i.e.

$$\iiint V_o dx dy dz = 0$$

where

$$V_o = \left(\sigma_x \varepsilon_x + \sigma_y \varepsilon_y + \sigma_z \varepsilon_z + \tau_{xy} \gamma_{xy} + \tau_{yz} \gamma_{yz} + \tau_{zx} \gamma_{zx} \right)$$

The strain energy per unit volume V_o is always positive for any combination of strains and stresses. Hence for the integral to be zero, V_o must vanish at all the points, giving all the stress components (or strain components) zero, for this case of zero body and surface forces.

Therefore,

$$(\sigma'_x - \sigma''_x) = (\sigma'_y, \sigma''_y) = (\sigma'_z - \sigma''_z) = 0$$

and

$$(\tau'_{xy} - \tau''_{xy}) = (\tau'_{yz} - \tau''_{yz}) = (\tau'_{zx} - \tau''_{zx}) = 0$$

This shows that the set $\sigma'_x, \sigma'_y, \sigma'_z, \ldots \tau\sigma'_{zx}$ is identical to the set $\sigma''_x, \sigma''_y, \sigma''_z, \ldots \tau''_{zx}$ and there is one and only one solution for the elastic problem.

4.9   Numerical Examples

Example 4.1 The following are the principal stresses at a point in a stressed material. Taking $E = 210\,\text{kN/mm}^2$ and $\nu = 0.3$, calculate the volumetric strain and Lame's constants.

$$\sigma_x = 200\,\text{N/mm}^2, \sigma_y = 150\,\text{N/mm}^2, \sigma_z = 120\,\text{N/mm}^2$$

Solution We have

$$\varepsilon_x = \frac{1}{E}\left[\sigma_x - \nu\left(\sigma_y + \sigma_z\right)\right]$$

$$= \frac{1}{210 \times 10^3}[200 - 0.3(150 + 120)]$$

$$\therefore \varepsilon_x = 5.67 \times 10^{-4}$$

$$\varepsilon_y = \frac{1}{E}\left[\sigma_y - \nu(\sigma_z + \sigma_x)\right]$$

$$= \frac{1}{210 \times 10^3}[150 - 0.3(120 + 200)]$$

$$\therefore \varepsilon_y = 2.57 \times 10^{-4}$$

$$\varepsilon_z = \frac{1}{E}\left[\sigma_z - \nu\left(\sigma_x + \sigma_y\right)\right]$$

$$= \frac{1}{210 \times 10^3}[120 - 0.3(200 + 150)]$$

$$\therefore \varepsilon_z = 7.14 \times 10^{-5}$$

$$\text{Volumetric strain} = \varepsilon_v = \left(\varepsilon_x + \varepsilon_y + \varepsilon_z\right)$$
$$= 5.67 \times 10^{-4} + 2.57 \times 10^{-4} + 7.14 \times 10^{-5}$$
$$\therefore \varepsilon_v = 8.954 \times 10^{-3}$$

To find Lame's constants
We have,

$$G = \frac{E}{2(1+v)}$$
$$G = \frac{210 \times 10^3}{2(1+0.3)}$$
$$\therefore G = 80.77 \times 10^3 \, \text{N/mm}^2$$

$$\lambda = \frac{G(2G-E)}{(E-3G)}$$
$$= \frac{80.77 \times 10^3 \left(2 \times 80.77 \times 10^3 - 210 \times 10^3\right)}{\left(210 \times 10^3 - 3 \times 80.77 \times 10^3\right)}$$
$$\therefore \lambda = 121.14 \times 10^3 \, \text{N/mm}^2$$

Example 4.2 The state of strain at a point is given by

$$\varepsilon_x = 0.001, \, \varepsilon_y = -0.003, \, \varepsilon_z = \gamma_{xy} = 0, \, \gamma_{xz} = -0.004, \, \gamma_{yz} = 0.001$$

Determine the stress tensor at this point. Take $E = 210 \times 10^6 \, \text{kN/m}^2$, Poisson's ratio = 0.28. Also find Lame's constant.

Solution We have

$$G = \frac{E}{2(1+v)}$$
$$= \frac{210 \times 10^6}{2(1+0.28)}$$
$$\therefore G = 82.03 \times 10^6 \, \text{kN/m}^2$$

But

$$\lambda = \frac{G(2G-E)}{(E-3G)}$$
$$= \frac{82.03 \times 10^6 \left(2 \times 82.03 \times 16^6 - 210 \times 10^6\right)}{\left(210 \times 10^6 - 3 \times 82.03 \times 10^6\right)}$$
$$\therefore \lambda = 104.42 \times 10^6 \, \text{kN/m}^2$$

Now,

$$\sigma_x = (2G + \lambda)\varepsilon_x + \lambda\left(\varepsilon_y + \varepsilon_z\right)$$
$$= (2 \times 82.03 + 104.42)10^6 \times 0.001 + 104.42 \times 10^6(-0.003 + 0)$$
$$\therefore \sigma_x = -44780 \, \text{kN/m}^2$$

or

$$\sigma_x = -44.78 MPa$$
$$\sigma_y = (2G + v)\varepsilon_y + \lambda(\varepsilon_z + \varepsilon_x)$$
$$= (2 \times 82.03 + 104.42) \times 10^6 \times (-0.003) + 104.42 \times 10^6(0 + 0.001)$$
$$\therefore \sigma_y = -701020 \, \text{kN/m}^2$$

or

$$\sigma_y = -701.02 \, \text{MPa}$$
$$\sigma_z = (2G + \lambda)\varepsilon_z + \lambda\left(\varepsilon_x + \varepsilon_y\right)$$
$$= (2 \times 82.03 + 104.42)10^6(0) + 104.42 \times 10^6(0.001 - 0.003)$$
$$\therefore \sigma_z = -208840 \, \text{kN/m}^2$$

or

$$\sigma_z = -208.84 \, \text{MPa}$$
$$\tau_{xy} = G\gamma_{xy}$$
$$= 82.03 \times 10^6 \times 0$$
$$\therefore \tau_{xy} = 0$$
$$\tau_{yz} = G\gamma_{yz} = 82.03 \times 10^6 \times 0.001 = 82030 \, \text{kN/m}^2$$

or

$$\tau_{yz} = 82.03 \, \text{MPa}$$
$$\tau_{xz} = G\gamma_{xz} = 82.03 \times 10^6 \times (-0.004) = -328120 \, \text{kN/m}^2$$
$$\text{or } \tau_{xz} = -328.12 \, \text{MPa}$$

$\therefore$ The stress tensor is given by

$$\sigma_{ij} = \begin{pmatrix} \sigma_x & \tau_{xy} & \tau_{xz} \\ \tau_{xy} & \sigma_y & \tau_{yz} \\ \tau_{xz} & \tau_{yz} & \sigma_z \end{pmatrix} = \begin{pmatrix} -44.78 & 0 & -328.12 \\ 0 & -701.02 & 82.03 \\ -328.12 & 82.03 & -208.84 \end{pmatrix}$$

Example 4.3 The stress tensor at a point is given as

$$\begin{pmatrix} 200 & 160 & -120 \\ 160 & -240 & 100 \\ -120 & 100 & 160 \end{pmatrix} \text{kN/m}^2$$

Determine the strain tensor at this point. Take $E = 210 \times 10^6 \text{ kN/m}^2$ and $v = 0.3$

Solution

$$\varepsilon_x = \frac{1}{E}[\sigma_x - v(\sigma_y + \sigma_z)]$$

$$= \frac{1}{210 \times 10^6}[200 - 0.3(-240 + 160)]$$

$$\therefore \varepsilon_x = 1.067 \times 10^{-6}$$

$$\varepsilon_y = \frac{1}{E}[\sigma_y - v(\sigma_z + \sigma_x)]$$

$$= \frac{1}{210 \times 10^6}[-240 - 0.3(160 + 200)]$$

$$\therefore \varepsilon_y = -1.657 \times 10^{-6}$$

$$\varepsilon_z = \frac{1}{E}[\sigma_z - v(\sigma_x + \sigma_y)]$$

$$= \frac{1}{210 \times 10^6}[160 - 0.3(200 - 240)]$$

$$\therefore \varepsilon_z = 0.82 \times 10^{-6}$$

Now,

$$G = \frac{E}{2(1+v)} = \frac{210 \times 10^6}{2(1+0.3)} = 80.77 \times 10^6 \text{ kN/m}^2$$

$$\tau_{xy} = G\gamma_{xy} = 80.77 \times 10^6 \times \gamma_{xy}$$

$$\therefore \gamma_{xy} = \frac{\tau_{xy}}{G} = \frac{160}{80.77 \times 10^6} = 1.981 \times 10^{-6}$$

$$\gamma_{yz} = \frac{\tau_{yz}}{G} = \frac{100}{80.77 \times 10^6} = 1.24 \times 10^{-6}$$

$$\gamma_{zx} = \frac{\tau_{zx}}{G} = \frac{-120}{80.77 \times 10^6} = -1.486 \times 10^{-6}$$

Therefore, the strain tensor at that point is given by

$$\varepsilon_{ij} = \begin{pmatrix} \varepsilon_x & \varepsilon_{xy} & \varepsilon_{xz} \\ \varepsilon_{xy} & \varepsilon_y & \varepsilon_{yz} \\ \varepsilon_{zx} & \varepsilon_{zy} & \varepsilon_z \end{pmatrix} = \begin{pmatrix} \varepsilon_x & \frac{\gamma_{xy}}{2} & \frac{\gamma_{xz}}{2} \\ \frac{\gamma_{xy}}{2} & \varepsilon_y & \frac{\gamma_{yz}}{2} \\ \frac{\gamma_{zx}}{2} & \frac{\gamma_{zy}}{2} & \varepsilon_z \end{pmatrix}$$

$$\therefore \varepsilon_{ij} = \begin{pmatrix} 1.067 & 0.9905 & -0.743 \\ 0.9905 & -1.657 & 0.62 \\ -0.743 & 0.62 & 0.82 \end{pmatrix} \times 10^{-6}$$

Example 4.4 A rectangular strain rosette gives the data as below.

$$\varepsilon_0 = 670 \text{ micrometres/m}$$
$$\varepsilon_{45} = 330 \text{ micrometres/m}$$
$$\varepsilon_{90} = 150 \text{ micrometres/m}$$

Find the principal stresses σ_1 and σ_2 if $E = 2 \times 10^5$ MPa, $v = 0.3$

Solution We have

$$\varepsilon_x = \varepsilon_0 = 670 \times 10^{-6}$$
$$\varepsilon_y = \varepsilon_{90} = 150 \times 10^{-6}$$

$$\gamma_{xy} = 2\varepsilon_{45} - (\varepsilon_0 + \varepsilon_{90}) = 2 \times 330 \times 10^{-6} - \left(670 \times 10^{-6} + 150 \times 10^{-6}\right)$$
$$\therefore \gamma_{xy} = -160 \times 10^{-6}$$

Now, the principal strains are given by

$$\varepsilon_{max} \text{ or } \varepsilon_{min} = \left(\tfrac{\varepsilon_x + \varepsilon_y}{2}\right) \pm \tfrac{1}{2}\sqrt{\left(\varepsilon_x - \varepsilon_y\right)^2 + \gamma_{xy}^2}$$

i.e. $\varepsilon_{max} \text{ or } \varepsilon_{min} = \left(\tfrac{670+150}{2}\right)10^{-6} \pm \tfrac{1}{2}\sqrt{\left[(670 - 150)10^{-6}\right]^2 + \left(-160 \times 10^{-6}\right)^2}$

$$\therefore \varepsilon_{max} \text{ or } \varepsilon_{min} = 410 \times 10^{-6} \pm 272.03 \times 10^{-6}$$
$$\therefore \varepsilon_{max} = \varepsilon_1 = 682.3 \times 10^{-6}$$
$$\varepsilon_{min} = \varepsilon_2 = 137.97 \times 10^{-6}$$

The principal stresses are determined by the following relations

$$\sigma_1 = \frac{(\varepsilon_1 + v\varepsilon_2)}{1 - v^2}.E = \frac{(682.03 + 0.3 \times 137.97)10^{-6}}{1 - (0.3)^2} \times 2 \times 10^5$$
$$\therefore \sigma_1 = 159 \text{ MPa}$$

Similarly,

$$\sigma_2 = \frac{(\varepsilon_2 + v\varepsilon_1)}{1 - v^2}.E = \frac{(137.97 + 0.3 \times 682.03)10^{-6}}{1 - (0.3)^2} \times 2 \times 10^5$$
$$\therefore \sigma_2 = 75.3 \text{ MPa}$$

4.10 Exercises

1. State and explain generalised Hooke's law
2. State and explain the following:

   (a) Saint-Venant's principle (b) Principle of superposition (c) Uniqueness theorem

3. The displacement field in a body is specified as $u = x^3 + 3$, $v = 3y^2z$, $w = x + 3z$. Determine the strain and stress components at a point whose coordinates are (1, 2, 3). Take $E = 2 \times 10^5 \text{N/mm}^2$ and $v = 0.32$.

4. A homogeneous and isotropic elastic solid with Young's modulus 200 GPa and Poisson's ratio 0.3 is subjected to a displacement field given by

$$u = (x^3y^2z - 4xy^2z^2)10^{-4}$$
$$v = (x^2y^2z^3 + 3xyz^2)10^{-4}$$
$$w = (2x^4 - 3y^2z^2 + xyz^4)10^{-4}$$

   Evaluate the stress components at a point whose coordinates are (2.5, 4.0, − 6.0). Express the above state of stress as the sum of its spherical and deviator components.

5. Calculate the principal strains and then the principal stresses at the point (2, 4) and the directions of principal stresses with respect to the x-axis for the following displacement field.

$$u = 4x^4 + 2x^2y^2 + x + 3$$
$$v = y^4 + 3x^2y + 1$$

6. Prove the following relationships:

   (i) $G = \frac{E}{2(1+v)}$ (ii) $\lambda = \frac{2Gv}{(1-2v)}$ (iii) $K = \frac{EG}{3(3G-E)}$

7. A triaxial state of principal stress acts on the faces of a unit cube. Show that these stresses will not produce a volume change if $v = 0.5$. The material is a linearly elastic isotropic material. If $v = 0.5$, show that the condition necessary for the volume to remain unchanged is $\sigma_1 + \sigma_2 + \sigma_3 = 0$.

8. A member made of isotropic bronze is subjected to a state of plane strain. Determine σ_z, ε_x, ε_y and γ_{xy}, if $\sigma_x = 90$ MPa, $\sigma_y = -50$ MPa, and $\tau_{xy} = 70$MPa.

   Take $E = 82.6$ GPa and $v = 0.35$.

9. A square plate with 800 mm sides parallel to the x and y-axes has a uniform thickness of 10 mm and is made of an isotropic steel ($E = 82.6$ GPa and $v = 0.29$). The plate is subjected to plane strain state. If $\sigma_x = \sigma_1 = 500$ MPa and $\varepsilon_x = 2\varepsilon_y$, determine the magnitude of $\sigma_y = \sigma_2$ and $\sigma_z = \sigma_3$, assuming linearly elastic conditions.

10. The components of a strain tensor at a point are given as:

$$\varepsilon_x = 0.01, \varepsilon_y = -0.03, \varepsilon_z = 0.02,$$
$$\gamma_{xy} = 0.01, \gamma_{yz} = -0.05, \gamma_{xz} = -0.02.$$

Determine the components of strain tensor at this point if $E = 200 \times 10^6$kPa and $G = 80 \times 10^6$kPa.

11. The stress components at a point are given as:

$$\sigma_x = 250\,\text{kPa}; \tau_{xy} = 210\,\text{kPa}$$
$$\sigma_y = -290\,\text{kPa}; \tau_{yz} = 150\,\text{kPa}$$
$$\sigma_z = 210\,\text{kPa}; \tau_{xz} = -170\,\text{kPa}$$

Determine the strain components at this point if $E = 210 \times 10^6$kPa and $\nu = 0.3$.

12. If a material has a Poisson's ratio of $\nu = 0.5$, show that

$$G = \frac{1}{3}E$$
$$K = \infty$$

Volumetric strain $(\varepsilon_v) = 0$.

13. A square plate 10 mm thick is subjected to a tensile stress $\sigma_x = 140$ MPa in one direction and compressive stress $\sigma_y = -140$ MPa in another direction. Find the change in volume of the plate if $E = 82.6$ GPa and $\nu = 0.29$.

14. A hard rubber block completely confined in the x-direction but free to expand in both the y and z directions is subjected to compressive stress $\sigma_y = -2$ MPa in y-direction as shown in figure below. Calculate the stress σ_x in the x-direction. What is the change in volume of the block if $E = 2$ MPa and $\nu = 0.5$ (Fig. 4.7).

Fig. 4.7 Rubber block subjected to compressive stress

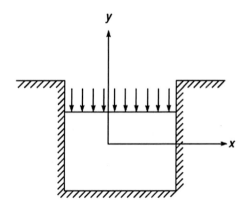

Chapter 5
Two-Dimensional Problems in Cartesian Co-ordinate System

5.1 Introduction

5.1.1 Plane Stress Problems

In many instances, the stress situation is simpler than that shown in Fig. 2.6. An example of practical interest is that of a thin plate which is being pulled by forces in the plane of the plate. Figure 5.1 shows a plate of constant thickness, t subjected to axial and shear stresses in the x-directions and y-directions only. These stresses are assumed to be uniformly distributed over the thickness t. The surface normal to the z-axis is stress free.

The state of stress at a given point will depend only on the three stress components such as

$$\sigma_x, \sigma_y \text{ and } \tau_{xy} = \tau_{yx} \qquad (5.1)$$

in which the stress components are functions of only x and y. This combination of stress components is called "plane stress" in the $x y$ plane. The stress–strain relations for plane stress are given by

$$\varepsilon_x = \frac{1}{E}\left(\sigma_x - v\sigma_y\right)$$
$$\varepsilon_y = \frac{1}{E}\left(\sigma_y - v\sigma_x\right)$$
$$\gamma_{xy} = \frac{\tau_{xy}}{G} \qquad (5.2)$$

and

$$\gamma_{xz} = \gamma_{yz} = 0, \quad \varepsilon_z = -\frac{v}{E}\left(\sigma_x + \sigma_y\right)$$

T. G. Sitharam and L. Govindaraju, *Theory of Elasticity*,
https://doi.org/10.1007/978-981-33-4650-5_5

Fig. 5.1 General case of
plane stress

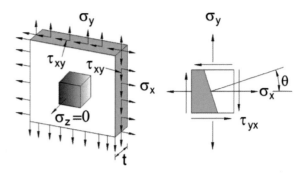

The constitutive relation for plane stress problems is given by

$$\left\{ \begin{array}{c} \sigma_x \\ \sigma_y \\ \tau_{xy} \end{array} \right\} = \frac{E}{(1 - v^2)} \begin{bmatrix} 1 & v & 0 \\ v & 1 & 0 \\ 0 & 0 & \left(\frac{1-v}{2}\right) \end{bmatrix} \left\{ \begin{array}{c} \varepsilon_x \\ \varepsilon_y \\ \gamma_{xy} \end{array} \right\} \tag{5.3}$$

5.1.2 Plane Strain Problems

Problems involving long bodies whose geometry and loading do not vary significantly
in the longitudinal direction are referred to as plane strain problems. Some examples
of practical importance, shown in Fig. 5.2, are a loaded semi-infinite half-space such
as a strip footing on a soil mass, a long cylinder; a tunnel; culvert; a laterally loaded
retaining wall; and a long earth dam. In these problems, the dependent variables can
be assumed to be functions of only the x and y-co-ordinates, provided a cross-section
is considered some distance away from the ends.

Hence the strain components will be

$$\varepsilon_x = \frac{\partial u}{\partial x}, \quad \varepsilon_y = \frac{\partial v}{\partial y}, \quad \gamma_{xy} = \frac{\partial u}{\partial y} + \frac{\partial v}{\partial x} \tag{5.4}$$

$$\varepsilon_z = \frac{\partial w}{\partial z} = 0, \quad \gamma_{xz} = \frac{\partial w}{\partial x} + \frac{\partial u}{\partial z} = 0, \quad \gamma_{yz} = \frac{\partial w}{\partial y} + \frac{\partial v}{\partial z} = 0 \tag{5.5}$$

Moreover, from the vanishing of ε_z, the stress σ_z can be expressed in terms of σ_x
and σ_y as

$$\sigma_z = v \left(\sigma_x + \sigma_y \right) \tag{5.6}$$

The constitutive law for elastic, isotropic material for plane strain problems is
given by

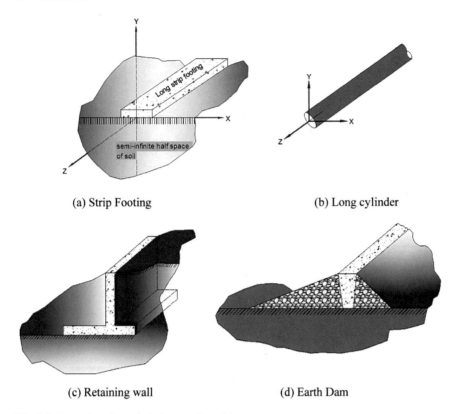

(a) Strip Footing

(b) Long cylinder

(c) Retaining wall

(d) Earth Dam

Fig. 5.2 Examples of practical plane strain problems

$$\left\{\begin{array}{c} \sigma_x \\ \sigma_y \\ \tau_{xy} \end{array}\right\} = \frac{E}{(1+v)(1-2v)} \left[\begin{array}{ccc} (1-v) & v & 0 \\ v & (1-v) & 0 \\ 0 & 0 & \left(\frac{1-2v}{2}\right) \end{array}\right] \left\{\begin{array}{c} \varepsilon_x \\ \varepsilon_y \\ \gamma_{xy} \end{array}\right\} \qquad (5.7)$$

5.2 Relationship Between Plane Stress and Plane Strain

5.2.1 Plane Stress Case

From the stress–strain relationship (Eq. 4.22), we have

$$\sigma_x = (2G + \lambda)\varepsilon_x + \lambda(\varepsilon_y + \varepsilon_z)$$

or

$$\sigma_x = 2G\,\varepsilon_x + \lambda\,\varepsilon_x + \lambda\big(\varepsilon_y + \varepsilon_z\big)$$

or

$$\sigma_x = \lambda\big(\varepsilon_x + \varepsilon_y + \varepsilon_z\big) + 2G\,\varepsilon_x$$

Similarly,

$$\sigma_y = \lambda\big(\varepsilon_x + \varepsilon_y + \varepsilon_z\big) + 2G\,\varepsilon_y$$
$$\sigma_z = \lambda\big(\varepsilon_x + \varepsilon_y + \varepsilon_z\big) + 2G\,\varepsilon_z = 0$$
$$\tau_{xy} = G\,\gamma_{xy} \quad \tau_{yz} = G\,\gamma_{yz} = 0 \quad \tau_{zx} = G\,\gamma_{zx} = 0$$

Denoting $\big(\varepsilon_x + \varepsilon_y + \varepsilon_z\big) = J_1 =$ First invariant of strain, then

$$\sigma_x = \lambda\,J_1 + 2\,G\varepsilon_x, \quad \sigma_y = \lambda\,J_1 + 2\,G\,\varepsilon_y, \quad \sigma_z = \lambda\,J_1 + 2\,G\,\varepsilon_z = 0 \qquad \text{(a)}$$

From, $\sigma_z = 0$, we get

$$\varepsilon_z = -\frac{\lambda}{(\lambda + 2G)}\big(\varepsilon_x + \varepsilon_y\big)$$

Using the above value of ε_z, the strain invariant J_1 becomes

$$J_1 = \frac{2G}{\lambda + 2G}\big(\varepsilon_x + \varepsilon_y\big) \qquad \text{(b)}$$

Substituting the value of J_1 in equation (a), we get

$$\sigma_x = \frac{2G\lambda}{\lambda + 2G}\big(\varepsilon_x + \varepsilon_y\big) + 2G\varepsilon_x$$
$$\sigma_y = \frac{2G\lambda}{\lambda + 2G}\big(\varepsilon_x + \varepsilon_y\big) + 2G\varepsilon_y$$

5.2.2   Plane Strain Case

Here $\varepsilon_z = 0$

$$\therefore \sigma_x = \lambda J_1 + 2G\varepsilon_x = \lambda\big(\varepsilon_x + \varepsilon_y\big) + 2G\varepsilon_x$$
$$\sigma_y = \lambda J_1 + 2G\varepsilon_y = \lambda\big(\varepsilon_x + \varepsilon_y\big) + 2G\varepsilon_y$$
$$\sigma_z = \lambda\,J_1 = \lambda\big(\varepsilon_x + \varepsilon_y\big)$$

If the equations for stress σ_x for plane strain and plane stress are compared, it can be observed that they are identical except for the comparison of coefficients of the term $(\varepsilon_x + \varepsilon_y)$.

i.e.

$$\sigma_x = \begin{cases} \lambda(\varepsilon_x + \varepsilon_y) + 2G\varepsilon_x & \text{plane strain} \\ \frac{2G\lambda}{\lambda+2G}(\varepsilon_x + \varepsilon_y) + 2G\varepsilon_x & \text{plane stress} \end{cases}$$

Since all the equations for stresses in plane stress and plane strain solutions are identical, the results from plane strain can be transformed into plane stress by replacing λ in plane strain case by $\frac{2G\lambda}{\lambda+2G}$ in plane stress case. This is equivalent to replacing $\frac{\nu}{1-\nu}$ in plane strain case by λ in plane stress case.

Similarly, a plane stress solution can be transformed into a plane strain solution by replacing $\frac{2G\lambda}{\lambda+2G}$ in plane stress case by λ in plane strain case. This is equivalent to replacing λ in plane stress case by $\frac{\nu}{1-\nu}$ in plane strain case.

5.3 Transformation of Compatibility Equation from Strain Components to Stress Components

5.3.1 Plane Stress Case

For two-dimensional problems, we have compatibility equation (from Eq. 3.38) as

$$\frac{\partial^2 \varepsilon_x}{\partial y^2} + \frac{\partial^2 \varepsilon_y}{\partial x^2} = \frac{\partial^2 \gamma_{xy}}{\partial x \, \partial y} \tag{5.8}$$

Further, stress–strain relations are given by (from Eq. 5.2)

$$\varepsilon_x = \frac{1}{E}\left(\sigma_x - \nu\sigma_y\right)$$

$$\varepsilon_y = \frac{1}{E}\left(\sigma_y - \nu\sigma_x\right)$$

$$\gamma_{xy} = \frac{\tau_{xy}}{G} \tag{5.9}$$

Substituting for $G = \frac{E}{2(1+\nu)}$ and Eq. (5.9) in Eq. (5.8), we get

$$\frac{\partial^2}{\partial^2 y}(\sigma_x - \nu\sigma_y) + \frac{\partial^2}{\partial x^2}(\sigma_y - \nu\sigma_x) = 2(1+\nu)\frac{\partial^2 \tau_{xy}}{\partial x \, \partial y} \tag{5.10}$$

Now, recalling equations of equilibrium (from Eqs. 2.31a and 2.31b)

$$\frac{\partial \sigma_x}{\partial x} + \frac{\partial \tau_{xy}}{\partial y} + F_x = 0 \tag{5.11}$$

$$\frac{\partial \sigma_y}{\partial y} + \frac{\partial \tau_{xy}}{\partial x} + F_y = 0 \tag{5.12}$$

Differentiating Eqs. (5.11) and (5.12) with respect to x and y, respectively, and adding, we get

$$\frac{\partial^2 \sigma_x}{\partial x^2} + \frac{\partial^2 \sigma_y}{\partial y^2} + 2\frac{\partial^2 \tau_{xy}}{\partial x\, \partial y} = -\left[\frac{\partial F_x}{\partial x} + \frac{\partial F_y}{\partial y}\right] \tag{5.13}$$

Substituting Eq. (5.13) in Eq. (5.10) we get

$$\left[\frac{\partial^2}{\partial x^2} + \frac{\partial^2}{\partial y^2}\right](\sigma_x + \sigma_y) = -(1 + v)\left[\frac{\partial F_x}{\partial x} + \frac{\partial F_y}{\partial y}\right] \tag{5.14}$$

If the body forces are constant or zero, then Eq. (5.14) results into Eq. (5.15). This Eq. (5.15) is the compatibility equation in terms of stresses for plane stress problems.

$$\left[\frac{\partial^2}{\partial x^2} + \frac{\partial^2}{\partial y^2}\right](\sigma_x + \sigma_y) = 0 \tag{5.15}$$

5.3.2 Plane Strain Case

The stress–strain relations for plane strain problems are given by

$$\varepsilon_x = \frac{1}{E}\{\sigma_x + v(\sigma_y + \sigma_z)\} \tag{5.16}$$

$$\varepsilon_y = \frac{1}{E}\{\sigma_y + v(\sigma_z + \sigma_x)\} \tag{5.17}$$

$$\gamma_{xy} = \frac{\tau_{xy}}{G} \tag{5.18}$$

$$\sigma_z = v(\sigma_x + \sigma_y) \quad \text{and} \quad G = \frac{E}{2(1 + v)} \tag{5.19}$$

The equilibrium equations, strain–displacement relations and the compatibility conditions are the same as for plane stress case also. Therefore substituting Eq. (5.16) to Eq. (5.19) in Eq. (5.8), we get

$$(1 - v)\left[\frac{\partial^2 \sigma_x}{\partial y^2} + \frac{\partial^2 \sigma_y}{\partial x^2}\right] - v\left[\frac{\partial^2 \sigma_y}{\partial y^2} + \frac{\partial^2 \sigma_x}{\partial x^2}\right] = 2\frac{\partial^2 \tau_{xy}}{\partial x \, \partial y} \tag{5.20}$$

Now differentiating equilibrium Eqs. (5.11) and (5.12) with respect to x and y, respectively, and adding the results as before and then substituting them in Eq. (5.20), we get

$$\left[\frac{\partial^2}{\partial x^2} + \frac{\partial^2}{\partial y^2}\right](\sigma_x + \sigma_y) = -\left(\frac{1}{1-v}\right)\left[\frac{\partial F_x}{\partial x} + \frac{\partial F_y}{\partial y}\right] \tag{5.21}$$

If the body forces are constant or zero, then Eq. (5.21) results into Eq. (5.22). This Eq. (5.22) is the compatibility equation in terms of stresses for plane strain problems.

$$\left[\frac{\partial^2}{\partial x^2} + \frac{\partial^2}{\partial y^2}\right](\sigma_x + \sigma_y) = 0 \tag{5.22}$$

It can be noted from Eqs. (5.15) and (5.22) that they are the same. Hence, if the body forces are constant or zero, the differential equations for plane stress will be same as for the plane strain. Moreover, it should be noted that neither the compatibility equations nor the equilibrium equations contain the elastic constants. Therefore, the stress distribution is same for all isotropic materials in two-dimensional state of stress.

5.4 Airy's Stress Function

Two-dimensional problems may be either formulated as plane stress or plane strain problems. The solutions to these problems may be obtained by several methods. But one among them is Airy's stress function method introduced by G.B. Airy. The relation between the stress function and the stresses is as follows.

$$\sigma_x = \frac{\partial^2 \phi}{\partial y^2}, \quad \sigma_y = \frac{\partial^2 \phi}{\partial x^2}, \quad \tau_{xy} = -\frac{\partial^2 \phi}{\partial x \partial y} \tag{5.23}$$

5.4.1 Stress Function for Plane Stress Case

In practice, the body forces are specified in terms of potential function, say Ω such that

$$F_x = -\frac{\partial \Omega}{\partial x} \quad \text{and} \quad F_y = -\frac{\partial \Omega}{\partial y} \tag{5.24}$$

Thus,

$$\sigma_x = \left(\frac{\partial^2 \phi}{\partial y^2} - \Omega\right) \quad \text{and} \quad \sigma_y = \left(\frac{\partial^2 \phi}{\partial x^2} - \Omega\right) \tag{5.25}$$

For two-dimensional problems with body forces, the equilibrium equations are given by

$$\frac{\partial \sigma_x}{\partial x} + \frac{\partial \tau_{xy}}{\partial y} + F_x = 0 \tag{5.26}$$

$$\frac{\partial \sigma_y}{\partial y} + \frac{\partial \tau_{xy}}{\partial x} + F_y = 0 \tag{5.27}$$

Substituting Eq. (5.24) in Eqs. (5.26) and (5.27), we get

$$\frac{\partial}{\partial x}(\sigma_x - \Omega) + \frac{\partial \tau_{xy}}{\partial y} = 0 \tag{5.28}$$

$$\frac{\partial}{\partial y}(\sigma_y - \Omega) + \frac{\partial \tau_{xy}}{\partial x} = 0 \tag{5.29}$$

But compatibility equation for strains is given by

$$\frac{\partial^2 \varepsilon_x}{\partial y^2} + \frac{\partial^2 \varepsilon_y}{\partial x^2} = \frac{\partial^2 \gamma_{xy}}{\partial x \partial y} \tag{5.30}$$

For plane stress problems, the stress–strain relations are

$$\varepsilon_x = \frac{1}{E}\left(\sigma_x - v\sigma_y\right)$$

$$\varepsilon_y = \frac{1}{E}\left(\sigma_y - v\sigma_x\right)$$

$$\varepsilon_z = -\frac{v}{E}\left(\sigma_x + \sigma_y\right)$$

$$\gamma_{xy} = \frac{\tau_{xy}}{G} = 2\frac{(1+v)}{E}\tau_{xy}$$

Substituting the above stress–strain relations in Eq. (5.30) and then simplifying, we get

$$\frac{\partial^2}{\partial y^2}\left[\sigma_x - v\sigma_y\right] + \frac{\partial^2}{\partial x^2}\left[\sigma_y - v\sigma_x\right] = 2(1+v)\frac{\partial^2 \tau_{xy}}{\partial x \partial y} \tag{5.31}$$

Substituting the values of $\sigma_x = \left(\frac{\partial^2 \phi}{\partial y^2} - \Omega\right)$, $\sigma_y = \left(\frac{\partial^2 \phi}{\partial x^2} - \Omega\right)$ and $\tau_{xy} = -\frac{\partial^2 \phi}{\partial x \partial y}$ in Eq. (5.31) and simplifying, one can get,

$$\frac{\partial^4 \phi}{\partial x^4} + \frac{\partial^4 \phi}{\partial y^4} + 2\frac{\partial^4 \phi}{\partial x^2 \partial y^2} + (1 - v)\left[\frac{\partial^2 \Omega}{\partial x^2} + \frac{\partial^2 \Omega}{\partial y^2}\right] = 0 \qquad (5.32)$$

Equation (5.32) is the governing differential equation for plane stress problems. If any function $\phi(x, y)$ can satisfy Eq. (5.32) and appropriate boundary conditions, then a plane stress problem can be solved. If the body forces are constant or zero, then Eq. (5.32) reduces to

$$\frac{\partial^4 \phi}{\partial x^4} + 2\frac{\partial^4 \phi}{\partial x^2 \partial y^2} + \frac{\partial^4 \phi}{\partial y^4} = 0 \qquad (5.33)$$

In general, $\nabla^4 \phi = 0$

where $\nabla^4 = \frac{\partial^4}{\partial x^4} + \frac{2\partial^4}{\partial x^2 \partial y^2} + \frac{\partial^4}{\partial y^4}$.

Equation (5.32) is known as "Biharmonic equation" for plane stress case.

5.4.2 Stress Function for Plane Strain Case

For plane-strain problem,

$$\varepsilon_z = 0 = \frac{1}{E}[\sigma_z - v(\sigma_x + \sigma_y)]$$
$$\sigma_z = v(\sigma_x + \sigma_y)] \qquad (5.34)$$

The stress-strain relations are

$$\varepsilon_x = \frac{1}{E}[\sigma_x - v(\sigma_y + \sigma_z)]$$
$$\varepsilon_y = \frac{1}{E}[\sigma_y - v(\sigma_x + \sigma_z)]$$
$$\varepsilon_z = \frac{1}{E}[\sigma_z - v(\sigma_x + \sigma_y)]$$
$$\gamma_{xy} = \frac{\tau_{xy}}{G} = \frac{2(1 + v)}{E}\tau_{xy} \qquad (5.35)$$

Substituting the value of σ_z in Eq. 5.35, we get

$$\varepsilon_x = \frac{1}{E}[\sigma_x - v\{\sigma_y + v(\sigma_x + \sigma_y)\}]$$

$$= \frac{1}{E}\left[(1 - v^2)\sigma_x - v(1 + v)\sigma_y\right]$$

$$= \frac{1}{E}(1 + v)\left[(1 - v)\sigma_x - v\sigma_y\right]$$

and

$$\varepsilon_x = \frac{1}{E}\left[\sigma_y - v(\sigma_x + \sigma_z)\right]$$

$$= \frac{1}{E}\left[\sigma_y - v\{\sigma_x + v(\sigma_x + \sigma_y)\}\right]$$

$$= \frac{1 + v}{E}\left[(1 - v)\sigma_y - v\sigma_x\right]$$

Now, rewriting the compatibility Eqs. (5.8) in terms of stress with the help of the above relations, we get

$$\frac{2(1 + v)}{E} \cdot \frac{\partial^2 \gamma_{xy}}{\partial x \partial y} = \frac{1 + v}{E} \cdot \frac{\partial^2}{\partial y^2}\left[(1 - v)\sigma_x - v\sigma_y\right] + \frac{1 + v}{E} \cdot \frac{\partial^2}{\partial x^2}\left[(1 - v)\sigma_x - v\sigma_x\right]$$

or

$$\frac{2\partial^2 \gamma_{xy}}{\partial x \partial y} = (1 - v)\frac{\partial^2 \sigma_x}{\partial y^2} - v\frac{\partial^2 \sigma_y}{\partial y^2} + (1 - v)\frac{\partial^2 \sigma_y}{\partial x^2} - v\frac{\partial^2 \sigma_x}{\partial x^2}$$

or

$$\frac{2\partial^2 \gamma_{xy}}{\partial x \partial y} = (1 - v)\left(\frac{\partial^2 \sigma_x}{\partial y^2} + \frac{\partial^2 \sigma_y}{\partial x^2}\right) - v\left(\frac{\partial^2 \sigma_y}{\partial y^2} + \frac{\partial^2 \sigma_x}{\partial x^2}\right) \qquad (5.36)$$

Now, differentiating Equation of (5.28) with respect to x and (5.29) with respect to y, we get

$$\frac{\partial^2 \sigma_x}{\partial x^2} - \frac{\partial^2 \Omega}{\partial x^2} + \frac{\partial^2 \tau_{xy}}{\partial x \partial y} = 0$$

$$\frac{\partial^2 \sigma_y}{\partial y^2} - \frac{\partial^2 \Omega}{\partial y^2} + \frac{\partial^2 \tau_{xy}}{\partial x \partial y} = 0$$

Substituting the above in Eq. (5.36), we get

$$2\frac{\partial^2 \tau_{xy}}{\partial x \partial y} = (1 - v)\left(\frac{\partial^2 \sigma_x}{\partial y^2} + \frac{\partial^2 \sigma_y}{\partial x^2}\right) - v\left(\frac{\partial^2 \Omega}{\partial x^2} + \frac{\partial^2 \Omega}{\partial y^2} - \frac{2\partial^2 \tau_{xy}}{\partial x \partial y}\right) = 0$$

or

$$2(1-v)\frac{\partial^4 \tau_{xy}}{\partial x \partial y} - (1-v)\left(\frac{\partial^2 \sigma_x}{\partial y^2} + \frac{\partial^2 \sigma_y}{\partial x^2}\right) + v\left(\frac{\partial^2 \Omega}{\partial x^2} + \frac{\partial^2 \Omega}{\partial y^2}\right) = 0$$

Now, substituting σ_x, σ_y and τ_{xy} in terms of stress function with body forces, we get

$$2(1-v)\frac{\partial^4 \phi}{\partial x^2 \partial y^2} + (1-v)\left[\left(\frac{\partial^2 \Omega}{\partial y^2} + \frac{\partial^4 \phi}{\partial y^4}\right) + \left(\frac{\partial^2 \Omega}{\partial x^2} + \frac{\partial^4 \phi}{\partial x^4}\right)\right]$$
$$- v\left(\frac{\partial^2 \Omega}{\partial x^2} + \frac{\partial^2 \Omega}{\partial y^2}\right) = 0$$

or

$$2(1-v)\frac{\partial^4 \phi}{\partial x^2 \partial y^2} + (1-v)\left[\left(\frac{\partial^2 \phi}{\partial x^4} + \frac{\partial^4 \phi}{\partial y^4}\right) + (1-2v)\left(\frac{\partial^2 \Omega}{\partial x^2} + \frac{\partial^2 \Omega}{\partial y^2}\right)\right] = 0$$

or

$$\frac{\partial^4 \phi}{\partial x^4} + 2\frac{\partial^4 \phi}{\partial x^2 \partial y^2} + \frac{\partial^4 \phi}{\partial y^4} + \frac{1-2v}{1-v}\left(\frac{\partial^2 \Omega}{\partial x^2} + \frac{\partial^2 \Omega}{\partial y^2}\right) = 0$$

or

$$\nabla^4 \phi + \left(\frac{1-2v}{1-v}\right)\nabla^2 \Omega = 0 \qquad (5.37)$$

If the body forces are constant or zero, Eq. (5.37) reduces to

$$\frac{\partial^4 \phi}{\partial x^4} + 2\frac{\partial^4 \phi}{\partial x^2 \partial y^2} + \frac{\partial^4 \phi}{\partial y^4} = 0 \qquad (5.38)$$

Equation (5.38) is identical to plane stress case. Therefore, Eq. (5.38) is same for the same geometry and surface forces, and hence, the stress distribution is same for plane stress and plane strain problems. Also, the Biharmonic equation does not involve any elastic constant, and thus, the stress distribution is independent of elastic constants in such cases.

Since the Biharmonic equation satisfies all the equilibrium and compatibility equations, a solution to this equation is also the solution for a two-dimensional problem. However, the solution in addition to satisfy the Biharmonic equation also has to satisfy the boundary conditions.

To solve the derived equations of elasticity, it is suggested to use polynomial functions, inverse functions or semi-inverse functions. The use of polynomial functions for solving two-dimensional problems is discussed in the next article. The inverse method requires examination of the assumed solutions with a view towards finding one which will satisfy the governing equations and the boundary conditions.

The semi-inverse method requires the assumption of a partial solution, formed by expressing stress, strain, displacement, or stress function in terms of known or undetermined coefficients. The governing equations are thus rendered more manageable.

5.5   Solution of Two-Dimensional Problems by the Use of Polynomials

The equation given by

$$\left(\frac{\partial^2 \phi}{\partial x^2} + \frac{\partial^2 \phi}{\partial y^2}\right) = \frac{\partial^4 \phi}{\partial x^4} + 2\frac{\partial^4 \phi}{\partial x^2 \partial y^2} + \frac{\partial^4 \phi}{\partial y^4} = 0 \tag{5.39}$$

will be satisfied by expressing Airy's function $\phi(x, y)$ in the form of homogeneous polynomials.

Usually, a polynomial is assumed for Airy's stress function, which satisfies the Biharmonic equation. Further, the assumed polynomial must also be such that the stress function satisfies boundary conditions. Any polynomial of less than or equal to third degree will satisfy the Biharmonic equation and is therefore a possible stress function. As the stresses are obtained from the stress function by its second derivatives, the second or higher order terms are essential in order to yield a nonzero stress solution of Eq. (5.39)

(a)   Polynomial of the First Degree

Let $\phi_1 = a_1 x + b_1 y$
   Now, the corresponding *stresses* are

$$\sigma_x = \frac{\partial^2 \phi_1}{\partial y^2} = 0, \sigma_y = \frac{\partial^2 \phi_1}{\partial x^2} = 0 \text{ and } \tau_{xy} = -\frac{\partial^2 \phi_1}{\partial x \partial y} 0$$

Therefore, this stress function gives a stress free body.

(b)   Polynomial of the Second Degree

Let $\phi_2 = \frac{a_2}{2} x^2 + b_2 xy + \frac{c_2}{2} y^2$
   The corresponding stresses are

$$\sigma_x = \frac{\partial^2 \phi_2}{\partial y^2} = c_2$$

Fig. 5.3 State of stresses

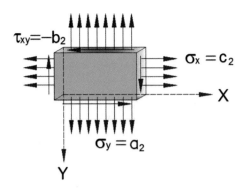

$$\sigma_y = \frac{\partial^2 \phi_2}{\partial x^2} = a_2$$

$$\tau_{xy} = -\frac{\partial^2 \phi}{\partial x \partial y} = -b_2$$

This shows that the above stress components do not depend upon the co-ordinates x and y; i.e. they are constant throughout the body representing a constant stress field. Thus, the stress function ϕ_2 represents a state of uniform tensions (or compressions) in two perpendicular directions accompanied with uniform shear, as shown in Fig. 5.3.

(c) Polynomial of the Third Degree

Let $\phi_3 = \frac{a_3}{6}x^3 + \frac{b_3}{2}x^2 y + \frac{c_3}{2}xy^2 + \frac{d_3}{6}y^3$

The corresponding stresses are

$$\sigma_x = \frac{\partial^2 \phi_3}{\partial y^2} = c_3 x + d_3 y$$

$$\sigma_y = \frac{\partial^2 \phi_3}{\partial x^2} = a_3 x + b_3 y$$

$$\tau_{xy} = -\frac{\partial^2 \phi_3}{\partial x \partial y} = -b_3 x - c_3 y$$

This stress function gives a linearly varying stress field. It should be noted that the magnitudes of the coefficients a_3, b_3, c_3 and d_3 are chosen freely since the expression for φ_3 is satisfied whatever values these coefficients have.

Now, if $a_3 = b_3 = c_3 = 0$ except d_3, we get from the stress components

$$\sigma_x = d_3 y$$
$$\sigma_y = 0 \text{ and } \tau_{xy} = 0$$

This corresponds to pure bending on the face perpendicular to the x-axis.

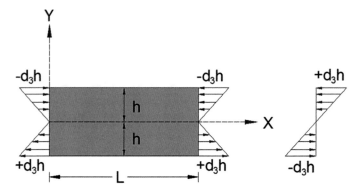

Fig. 5.4 Variation of stresses

$$\therefore \quad \text{At } y = -h, \sigma_x = -d_3 h$$

and

$$\text{At } y = +h, \sigma_x = +d_3 h$$

The variation of σ_x with y is linear as shown in Fig. 5.4.
Similarly, if all the coefficients except b_3 are zero, then we get

$$\sigma_x = 0$$
$$\sigma_y = b_3 y$$
$$\tau_{xy} = -b_3 x$$

The stresses represented by the above stress field will vary as shown in Fig. 5.5.

Fig. 5.5 Variation of
stresses

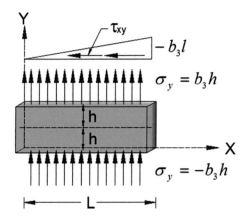

In Fig. 5.5, the stress σ_y is constant with x (i.e. constant along the span L of the beam), but varies with y at a particular section. At $y = +h$, $\sigma_y = b_3h$ (i.e. tensile), while at $y = -h$, $\sigma_y = -b_3h$ (i.e. compressive). σ_x is zero throughout. Shear stress τ_{xy} is zero at $x = 0$ and is equal to $-b_3L$ at $x = L$. At any other section, the shear stress is proportional to x.

(d) Polynomial of the Fourth Degree

Let $\phi_4 = \frac{a_4}{24}x^4 + \frac{b_4}{6}x^3y + \frac{c_4}{2}x^2y^2 + \frac{d_4}{6}xy^3 + \frac{e_4}{24}y^4$
The corresponding stresses are given by

$$\sigma_x = c_4x^2 + d_4xy + \frac{e_4}{2}y^2$$

$$\sigma_y = \frac{a_4}{2}x^2 + b_4xy + c_4y^2$$

$$\tau_{xy} = -\left(\frac{b_4}{2}\right)x^2 - 2c_4xy - \left(\frac{d_4}{2}\right)y^2$$

Now, taking all coefficients except d_4 equal to zero, we find

$$\sigma_x = d_4xy, \quad \sigma_y = 0, \quad \tau_{xy} = -\frac{d_4}{2}y^2$$

Assuming d_4 positive, the forces acting on the beam are shown in Fig. 5.6. Now, couple formed by stress σ_x is

$$M = \int_{-h}^{+h} (d_4xy)y\,dy$$

$$= \frac{2}{3}d_4Lh^3$$

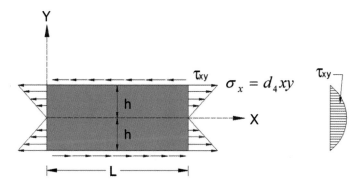

Fig. 5.6 Stresses acting on the beam

The couple due to the shearing stresses τ_{xy} is

$$M = \frac{d_4}{2}h^2 L2h + L\int_{-h}^{+h}\frac{d_4}{2}y^2 dy$$

$$= -\frac{2}{3}d_4 Lh^3$$

Therefore, the net couple on the plate is zero.

From the above consideration, it is to be noted that for stress function in the form of polynomials of second and third degree, all constants are arbitrary since the Biharmonic equation is satisfied (i.e. $\nabla^4 \phi = 0$) regardless of the values of the constants. However, for polynomials of higher order, the Biharmonic equation is satisfied under certain conditions and all the constants are not arbitrary. As an example, the fourth degree polynomial equation considered here is satisfied when $e_4 = -(a_4 + 4\,c_4)$.

(e) Polynomial of the Fifth Degree

Let

$$\phi_5 = \frac{a_5}{5(4)}x^5 + \frac{b_5}{4(3)}x^4 y + \frac{c_5}{3(2)}x^3 y^2 + \frac{d_5}{3(2)}x^2 y^3 + \frac{e_5}{4(3)}xy^4 + \frac{f_5}{5(4)}y^5$$

The corresponding stress components are given by

$$\sigma_x = \frac{\partial^2 \phi_5}{\partial y^2} = \frac{c_5}{3}x^3 + d_5 x^2 y - (2c_5 + 3a_5)xy^2 - \frac{1}{3}(b_5 + 2d_5)y^3$$

$$\sigma_y = \frac{\partial^2 \phi_5}{\partial x^2} = a_5 x^3 + b_5 x^2 y + c_5 xy^2 + \frac{d_5}{3}y^3$$

$$\tau_{xy} = -\frac{\partial^2 \phi_5}{\partial x \partial y} = -\frac{1}{3}b_5 x^3 - c_5 x^2 y - d_5 xy^2 + \frac{1}{3}(2c_5 + 3a_5)y^3$$

Here the coefficients a_5, b_5, c_5, d_5 are arbitrary, and in adjusting them we obtain solutions for various loading conditions of the beam.

Now, if all coefficients, except d_5, equal to zero, we find

$$\sigma_x = d_5\left(x^2 y - \frac{2}{3}y^3\right)$$

$$\sigma_y = \frac{1}{3}d_5 y^3$$

$$\tau_{xy} = -d_5 xy^2$$

Case (i) The normal forces are uniformly distributed along the longitudinal sides of the beam.

Case (ii) Along the side $x = L$, the normal forces consist of two parts, one following a linear law and the other following the law of a cubic parabola. The shearing forces are proportional to x on the longitudinal sides of the beam and follow a parabolic law along the side $x = L$.

The distribution of the stresses for ***Case (i)*** and ***Case (ii)*** is shown in Fig. 5.7.

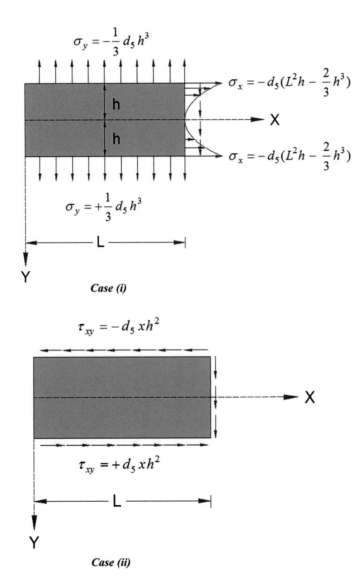

Fig. 5.7 Distribution of forces on the beam

5.6   Pure Bending of a Beam

Consider a rectangular beam, length L, width $2b$, depth $2\,h$, subjected to a pure couple M along its length as shown in Fig. 5.8.

Consider a second-order polynomial such that its any term gives only a constant state of stress. Therefore,

$$\phi = a_2\frac{x^2}{2} + b_2xy + \frac{c_2y^2}{2} \tag{5.40}$$

By definition,

$$\sigma_x = \frac{\partial^2\phi}{\partial y^2}, \quad \sigma_y = \frac{\partial^2\phi}{\partial x^2}, \quad \tau_{xy} = -\left(\frac{\partial^2\phi}{\partial x\partial y}\right)$$

$\therefore$ Differentiating the function, we get

$$\sigma_x = \frac{\partial^2\phi}{\partial y^2} = c_2, \quad \sigma_y = \frac{\partial^2\phi}{\partial x^2} = a_2 \quad \text{and} \quad \tau_{xy} = -\left(\frac{\partial^2\phi}{\partial x\partial y}\right) = -b_2$$

Considering the plane stress case,

$$\sigma_z = \tau_{xz} = \tau_{yz} = 0$$

Boundary Conditions

(a) At $y = \pm h$, $\sigma_y = 0$
(b) At $y = \pm h$, $\tau_{xy} = 0$
(c) At $x = $ any value,

$$2b\int_{-h}^{+h} \sigma_x y\,dy = \text{bending moment(M)} = \text{constant}$$

Fig. 5.8 Beam under pure bending

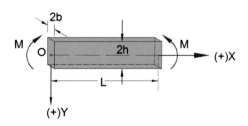

$$\therefore 2bx \int\limits_{-h}^{+h} c_2 y \, dy = 2bc_2 x \left[\frac{y^2}{2}\right]_{-h}^{+h} = 0$$

Therefore, this clearly does not fit the problem of pure bending.
Now, consider a third-order equation

$$\phi = \frac{a_3 x^3}{6} + \frac{b_3}{2} x^2 y + \frac{c_3 x y^2}{2} + \frac{d_3 y^3}{6}$$

Now,

$$\sigma_x = \frac{\partial^2 \phi}{\partial y^2} = c_3 x + d_3 y \tag{a}$$

$$\sigma_y = a_3 x + b_3 y \tag{b}$$

$$\tau_{xy} = -b_3 x - c_3 y \tag{c}$$

From (b) and boundary condition (a) above,

$$0 = a_3 x \pm b_3 a \text{ for any value of } x$$
$$\therefore \ a_3 = b_3 = 0$$

From (c) and the above boundary condition (b),

$$0 = -b_3 x \pm c_3 a \text{ for any value of } x$$

therefore $c_3 = 0$

hence,

$$\sigma_x = d_3 y$$
$$\sigma_y = 0$$
$$\tau_{xy} = 0$$

Obviously, Biharmonic equation is also satisfied.

i.e. $\frac{\partial^4 \phi}{\partial x^4} + 2\frac{\partial^4 \phi}{\partial x^2 \partial y^2} + \frac{\partial^4 \phi}{\partial y^4} = 0$.

Now,

$$\text{Bending moment} = M = 2b \int\limits_{-h}^{+h} \sigma_x y \, dy$$

$$\text{i.e.} \quad M = 2b \int_{-h}^{+h} d_3 y^2 dy$$

$$= 2bd_3 \int_{-h}^{+h} y^2 dy$$

$$= 2bd_3 \left[\frac{y^3}{3} \right]_{-h}^{+h}$$

$$M = 4bd_3 \frac{h^3}{3}$$

Or

$$d_3 = \frac{3M}{4bh^3}$$

$$d_3 = \frac{M}{I}$$

where $I = \frac{4h^3 b}{3}$.

Therefore,

$$\sigma_x = \frac{M}{I} y \tag{5.41}$$

From the above, it can be indicated that the simple theory of bending gives the same solution.

5.7 Bending of a Narrow Cantilever Beam Subjected to End Load

Consider a cantilever beam having a narrow rectangular cross section of unit width and depth 2 h as shown in Fig. 5.9. Let P be a force applied at its free end as shown.

Let us a adopt fourth-degree polynomial function

i.e.,

$$\phi = \frac{a_4}{24} x^4 + \frac{b_4}{6} x^3 y + \frac{c_4}{2} x^2 y^2 + \frac{d_4}{6} x y^3 + \frac{e_4}{24} y^4$$

Here,

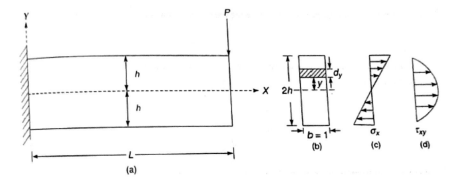

Fig. 5.9 a Cantilever beam with point load ar end, **b** cross section of beam, **c** variation of bending stress, **d** variation of shear stress

$$\frac{\partial^4 \phi}{\partial x^4} = a_4, \quad \frac{\partial^4 \phi}{\partial y^4} = e_4 \quad \text{and} \quad \frac{2\partial^4 \phi}{\partial x^2 \partial y^2} = 4c_4$$

Biharmonic equation is given by

$$\frac{\partial^4 \phi}{\partial x^4} + \frac{2\partial^4 \phi}{\partial x^2 \partial y^2} + \frac{\partial^4 \phi}{\partial y^4} = 0 \tag{5.42}$$

Substituting the above values in Eq. (5.42), we get

$$a_4 + 4c_4 + e_4 \neq 0$$

i.e.,

$$a_4 + 4c_4 + e_4 = 0$$

Hence,

$$e_4 = -(a_4 + 4c_4)$$

Now the stress components ar given by

$$\sigma_x = \frac{\partial^2 \phi}{\partial y^2} = c_4 x^2 + d_4 xy + \frac{e_4}{2} y^2$$

or

$$\sigma_x = c_4 x^2 + d_4 xy - \frac{(a_4 + 4c_4)}{2} y^2$$

$$= c_4 x^2 + d_4 xy - \frac{a_4}{2} y^2 - 2c_4 y^2$$

$$\sigma_y = \frac{\partial^2 \phi}{\partial x^2} = \frac{a_4}{2} x^2 + b_4 xy + c_4 y^2$$

and

$$\tau_{xy} = -\frac{\partial^2 \phi}{\partial x \partial y} = \frac{b_4}{2} x^2 - 2c_4 xy - \frac{d_4}{2} y^2$$

For the given problem, the upper and lower edges ar free from load.

i.e. at

$$y = \pm h, \quad \sigma_y = 0$$

and at

$$y = \pm h, \quad \tau_{xy} = 0$$

Further, the shearing forces, having a resultant P are distributed along the end $x = 0$.

Now, taking the consultant $c_4 = a_4 = b_4 = 0$ and $d_4 \neq 0$,

then

$$\sigma_x = d_4 xy \qquad\qquad (5.43)$$

$$\sigma_y = 0 \qquad\qquad (5.44)$$

and

$$\tau_{xy} = -\frac{d_4}{2} y^2 \qquad\qquad (5.45)$$

But the beam is subjected to a constant shear force 'P' resulting on constant shear stress distribution. Hence, superimposing the pure shear component (from second degree polynomial), i.e., $\tau_{xy} = -b_2$ in Eq. (5.45), the resulting final stresses are given by

$$\sigma_x = d_4 xy$$
$$\sigma_y = 0$$
$$\tau_{xy} = -b_2 - \frac{d_4}{2} y^2$$

Boundary conditions

At

$$y = \pm h, \quad \sigma_y = 0$$

At

$$y = \pm h, \quad \tau_{xy} = 0$$

Hence,

$$0 = -b_2 \pm \frac{d_4}{2} h^2$$

or

$$d_4 = -2 \frac{b_2}{h^2}$$

Now consider an elemental strip of thickness 'dy' at a distance 'y' above centroidal axis as shown in Fig. 5.9b.

Therefore, $\int_{-h}^{+h} \tau_{xy} \cdot b \, dy = -P$ (−ve sign denote load acting downwards).

i.e.,

$$-P = \int_{-h}^{+h} \left(-b_2 - \frac{d_4}{2} y^2 \right) \cdot 1 \cdot dy$$

But

$$d_4 = -2 \frac{b_2}{h^2}$$

Hence,

$$-P = \int\limits_{-h}^{+h} -b_2 \, dy + \int\limits_{-h}^{+h} \left(-2\frac{b_2}{h^2}\right) \frac{y^2}{2} dy$$

Simplifying, we get

$$b_2 = \frac{3P}{4h}$$

Therefore, expressions for σ_x, σ_y and τ_{xy} can be written as

$$\sigma_x = -\frac{2b_2}{h^2} xy = \left(-\frac{3}{2} \cdot \frac{P}{h^3}\right) xy$$

or

$$\sigma_x = -\frac{P}{I} xy$$

Here I = Moment of inertia of c/s of beam = $\frac{2}{3}h^3$

$$\sigma_y = 0$$

and

$$\tau_{xy} = -\frac{3P}{4h} + \frac{3P}{4h^3} \cdot y^2$$

or

$$\tau_{xy} = -\frac{3P}{4h}\left(1 - \frac{y^2}{h^2}\right)$$

$$= -\frac{3P}{4h^3}\left(h^2 - y^2\right)$$

$$\tau_{xy} = -\frac{P}{2I}\left(h^2 - y^2\right)$$

This distribution of bending stress and shear stress is shown in Fig. 5.9c, d respectively.

5.8 Bending of a Simply Supported Beam by a Distributed Loading (Udl)

Consider a beam of rectangular cross-section having unit width, supported at the ends and subjected to a uniformly distributed load of intensity q as shown in Fig. 5.10.

It is to be noted that the bending moment is a maximum at position $x = 0$ and decreases with a change in x in either the positive or the negative direction. This is possible only if the stress function contains even functions of x. Also, it should be noted that σ_y varies from zero at $y = -h$ to a maximum value of $-q$ at $y = +h$. Hence, the stress function must contain odd functions of y.

Now, consider a polynomial of second degree with

$$b_2 = c_2 = 0$$

$$\therefore \phi_2 = \frac{a_2}{2} x^2$$

a polynomial of third degree with $a_3 = c_3 = 0$

$$\therefore \phi_3 = \frac{b_3}{2} x^2 y + \frac{d_3}{6} y^3$$

and a polynomial of fifth degree with $a_5 = b_5 = c_5 = e_5 = 0$

$$\therefore \phi_5 = \frac{d_5}{6} x^2 y^3 - \frac{d_5}{30} y^5 \quad \left[\because f_5 = -\frac{2}{3} d_5 \right]$$

$$\therefore \phi = \phi_2 + \phi_3 + \phi_5$$

or

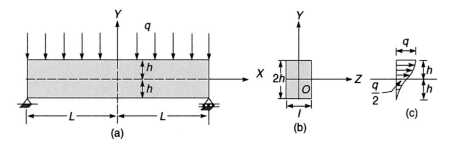

(a) (b) (c)

Fig. 5.10 Beam subjected to Uniform load

$$\phi = \frac{a_2}{2}x^2 + \frac{b_3}{2}x^2y + \frac{d_3}{6}y^3 + \frac{d_5}{6}x^2y^3 - \frac{d_5}{30}y^5 \qquad (5.46a)$$

Now, by definition,

$$\sigma_x = \frac{\partial^2 \phi}{\partial y^2} = d_3 y + d_5\left(x^2 y - \frac{2}{3}y^3\right) \qquad (5.46b)$$

$$\sigma_y = \frac{\partial^2 \phi}{\partial x^2} = a_2 + b_3 y + \frac{d_5}{3}y^3 \qquad (5.46c)$$

$$\tau_{xy} = -b_3 x - d_5 xy^2 \qquad (5.46d)$$

The following boundary conditions must be satisfied.

(i) $\left(\tau_{xy}\right)_{y=\pm h} = 0$

(ii) $\left(\sigma_y\right)_{y=+h} = 0$

(iii) $\left(\sigma_y\right)_{y=-h} = -q$

(iv) $\int\limits_{-h}^{+h} (\sigma_x)_{x=\pm L} dy = 0$

(v) $\int\limits_{-h}^{+h} \left(\tau_{xy}\right)_{x=\pm L} dy = \pm q L$

(vi) $\int\limits_{-h}^{+h} (\sigma_x)_{x=\pm L}\, y\, dy = 0$

The first three conditions when substituted in Eqs. (5.46c) and (5.46d) give

$$-b_3 - d_5 h^2 = 0$$

$$a_2 + b_3 h + \frac{d_5}{3}h^3 = 0$$

$$a_2 - b_3 h - \frac{d_5}{3}h^3 = -q$$

which gives on solving

$$a_2 = -\frac{q}{2}, \quad b_3 = \frac{3q}{4h}, \quad d_5 = -\frac{3q}{4h^3}$$

Now, from condition (vi), we have

$$\int\limits_{-h}^{+h}\left[d_3 y + d_5\left(x^2 y - \frac{2}{3}y^3\right)\right] y\, dy = 0$$

Simplifying,

$$d_3 = -d_5\left(L^2 - \frac{2}{5}h^2\right)$$

$$= \frac{3q}{4h^3}\left(\frac{L^2}{h^2} - \frac{2}{5}\right)$$

$$\therefore \sigma_x = \frac{3q}{4h^3}\left(\frac{L^2}{h^2} - \frac{2}{5}\right)y - \frac{3q}{4h^3}\left(x^2y - \frac{2}{3}y^3\right)$$

$$\sigma_y = -\left(\frac{q}{2}\right) + \frac{3q}{4h}y - \frac{q}{4h^3}y^3$$

$$\tau_{xy} = -\left(\frac{3q}{4h}\right)x + \frac{3q}{4h^3}xy^2$$

Now, $I = \frac{1\times(2h)^3}{12} = \frac{8h^3}{12} = \frac{2}{3}h^3$

where I = Moment of inertia of the unit width beam.

$$\therefore \sigma_x = \frac{q}{2I}(L^2 - x^2)y + \frac{q}{I}\left(\frac{y^3}{3} - \frac{h^2y}{5}\right) \tag{5.47}$$

$$\sigma_y = -\left(\frac{q}{2I}\right)\left(\frac{y^3}{3} - h^2y + \frac{2}{3}h^3\right) \tag{5.48}$$

$$\tau_{xy} = -\left(\frac{q}{2I}\right)(h^2 - y^2)x \tag{5.49}$$

Here the first term of the Eq. (5.47) represents the stresses given by the elementary theory of bending. But the second term of the equation gives necessary correction. It is to be noted that the term for correction does not depend on x and is small when compared to the maximum bending stress, provided the span of the beam is large in comparison with the depth. Therefore, for such beams, the usual elementary theory of bending provides a sufficiently accurate value for the stresses σ_x.

Rearranging the expressions for σ_x, we get

$$\sigma_x = \frac{q}{2I}(L_2 - x^2)y + \frac{q}{2I}\left(\frac{2}{3}y^3 - \frac{2}{5}h^2y\right) \tag{5.50}$$

The expressions (5.44) or (5.47) is an exact solution only if at the ends of the beam, i.e., $x = \pm L$ the normal stresses are distributed according to the law

$$T_x = \pm\frac{q}{2I}\left(\frac{2}{3}y^3 - \frac{2}{5}h^2y\right)$$

or

$$T_x = \pm \frac{3q}{4h^3} \left(\frac{2}{3} y^3 - \frac{2}{5} h^2 y \right)$$

i.e., if the normal stresses at the ends are the same as σ_x for $x = \pm L$ from Eq. (5.50) these stresses have zero resultant force and zero resultant moment. Therefore, from Saint-Venant's principle it can be concluded that their effect on the stresses far away from the ends for example at distances larger than depth of the beam can be neglected. Hence, solution for stresses at such points is accurate when no such stresses T, are applied.

The main differences between the exact solution Eq. (5.50) and the approximate solution given by the first term of Eq. (5.50) is due to the assumption that the longitudinal fibers of the beam are in a condition of simple tension while tension while deriving the approximate solution. From the equation for σ_y Eq. (5.48), it can be seen that there are compressive stresses between the fibers and these are responsible for the correction indicated by the second term of the Eq. (5.50). The variation of compressive stress σ_y over the depth of the beam is shown in Fig. 5.10c.

Further, the shearing stress distribution τ_{xy} given by the Eq. (5.49) over the cross section of the beam coincides with the solution given by elementary theory of bending.

5.9 Numerical Examples

Example 5.1
Show that for a simply supported beam, length $2L$, depth $2h$ and unit width, loaded by a concentrated load at mid span, the stress function satisfying the loading condition is $\phi = \frac{b}{6} xy^2 + cxy$. Treat the concentrated load as a shear stress suitably distributed to suit this function, so that $\int_{-h}^{+h} \tau_{xy} dy = -\left(\frac{W}{2}\right)$ on each half-length of the beam. Also find the stresses in the beam (Fig. 5.11).

Solution The stress components obtained from the stress function are

$$\sigma_x = \frac{\partial^2 \phi}{\partial y^2} = bxy$$

$$\sigma_y = \frac{\partial^2 \phi}{\partial x^2} = 0$$

$$\tau_{xy} = -\frac{\partial^2 \phi}{\partial x \partial y} = -\left(\frac{by^2}{2}\right) + c$$

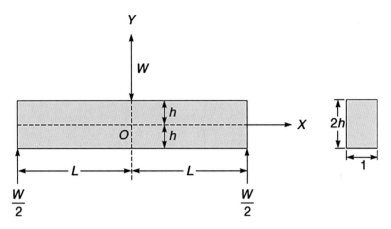

Fig. 5.11 Simply supported beam

Boundary conditions are

(i) $\sigma_y = 0$ for $y = \pm h$

(ii) $\tau_{xy} = 0$ for $y = \pm h$

(iii) $-\displaystyle\int_{-h}^{+h} \tau_{xy} \mathrm{d}y = \dfrac{W}{2}$ for $x = \pm L$

(iv) $\displaystyle\int_{-h}^{+h} \sigma_x \mathrm{d}y = 0$ for $x = \pm L$

(v) $\displaystyle\int_{-h}^{+h} \sigma_x y\, \mathrm{d}y = 0$ for $x = \pm L$

Now,

Condition (i)
This condition is satisfied since $\sigma_y = 0$.

Condition (ii)

$$0 = -\left(\frac{ba^2}{2}\right) + c$$

$$\therefore c = \frac{bh^2}{2}$$

Condition (iii)

$$\frac{W}{2} = -\int_{-h}^{+h} \tau_{xy}\, dy$$

$$\frac{W}{2} = -\int_{-h}^{+h} \left(-\frac{b}{2}y^2 - c\right) dy$$

$$\frac{W}{2} = \int_{-h}^{+h} \left(\frac{b}{2}y^2 + c\right) dy$$

$$= \frac{b}{3}h^3 + 2ch$$

Substituting for the value of c, we get

$$\frac{W}{2} = -\frac{2}{3}bh^3$$

Hence,

$$b = -\left(\frac{3W}{4h^3}\right) \quad \text{and} \quad c = -\left(\frac{3W}{8h}\right)$$

Condition (iv)

Now,

$$\sigma_x = -\left(\frac{3W}{4h^3}\right)xy$$

Hence,

$$M = \int_{-h}^{+h} -\left(\frac{3W}{4h^3}\right)xy^2\, dy = 0$$

Therefore,

$$M = -\frac{W\,x}{2}$$

Hence, stress components are

$$\sigma_x = -\left(\frac{3W}{4h^3}\right)xy$$

$$\sigma_y = 0$$

$$\tau_{xy} = -\frac{3W}{8h}\left[1 - \frac{y^2}{h^2}\right]$$

Example 5.2

Given the stress function $\phi = \left(\frac{H}{\pi}\right)z \tan^{-1}\left(\frac{x}{z}\right)$. Determine whether stress function ϕ is admissible. If so determine the stresses.

Solution For the stress function ϕ to be admissible, it has to satisfy Biharmonic equation. Biharmonic equation is given by

$$\frac{\partial^4\phi}{\partial x^4} + 2\frac{\partial^4\phi}{\partial x^2 \partial z^2} + \frac{\partial^4\phi}{\partial z^4} = 0 \tag{i}$$

Now,

$$\frac{\partial\phi}{\partial z} = \frac{H}{\pi}\left[-\left(\frac{xz}{x^2 + z^2}\right) + \tan^{-1}\left(\frac{x}{z}\right)\right]$$

$$\frac{\partial^2\phi}{\partial z^2} = \left(\frac{H}{\pi}\right)\frac{1}{(x^2 + z^2)^2}[2xz^2 - xz^2 - x^3 - xz^2 - x^3]$$

$$\therefore \frac{\partial^2\phi}{\partial z^2} = -\left(\frac{2H}{\pi}\right)\left[\frac{x^3}{(x^2 + z^2)^2}\right]$$

Also,

$$\frac{\partial^3\phi}{\partial z^3} = \frac{H}{\pi}\left[\frac{8x^3 z}{(x^2 + z^2)^3}\right]$$

$$\frac{\partial^4\phi}{\partial z^4} = \frac{H}{\pi}\left[\frac{8x^5 - 40x^3 z^2}{(x^2 + z^2)^4}\right]$$

$$\frac{\partial^3\phi}{\partial z^2 \partial x} = -\left(\frac{2H}{\pi}\right)\left[\frac{3x^3 z^2 - x^4}{(x^2 + z^2)^3}\right]$$

$$\frac{\partial^4\phi}{\partial z^2 \partial x^2} = \frac{H}{\pi}\left[\frac{64x^3 z^2 - 24xz^4 - 8x^5}{(x^2 + z^2)^4}\right]$$

Similarly,

$$\frac{\partial \phi}{\partial x} = \frac{H}{\pi}\left[\frac{z^2}{(x^2+z^2)}\right]$$

$$\frac{\partial^2 \phi}{\partial x^2} = -\left(\frac{2H}{\pi}\right)\left[\frac{xz^2}{(x^2+z^2)^2}\right]$$

$$\frac{\partial^3 \phi}{\partial x^3} = \frac{2H}{\pi}z^2\left[\frac{(3x^2-z^2)}{(x^2+z^2)^3}\right]$$

$$\frac{\partial^4 \phi}{\partial x^4} = \frac{H}{\pi}\left[\frac{24xz^4 - 24x^3z^2}{(x^2-z^2)^4}\right]$$

Substituting the above values in (i), we get

$$\frac{4}{\pi}\frac{1}{(x^2+z^2)^4}\left[24xz^4 - 24x^3z^2 + 64x^3z^2 - 24xz^4 - 8x^5 + 8x^5 - 40x^3z^2\right] = 0$$

Hence, the given stress function is admissible.
Therefore, the stresses are

$$\sigma_x = \frac{\partial^2 \phi}{\partial z^2} = -\left(\frac{24}{\pi}\right)\left[\frac{x^3}{(x^2+z^2)^2}\right]$$

$$\sigma_y = \frac{\partial^2 \phi}{\partial x^2} = -\left(\frac{24}{\pi}\right)\left[\frac{x^2}{(x^2+z^2)^2}\right]$$

and

$$\tau_{xy} = \frac{\partial^2 \phi}{\partial x \partial z} = -\left(\frac{24}{\pi}\right)\left[\frac{x^2 z}{(x^2+z^2)^2}\right]$$

Example 5.3
Given the stress function $\phi = -(\frac{F}{h^3})xy^2(3h-2y)$. Determine the stress components and sketch their variations in a region included in y = 0, y = h, x = 0, on the side x positive.

Solution The given stress function may be written as

$$\phi = -\left(\frac{3F}{h^2}\right)xy^2 + \left(\frac{2F}{h^3}\right)xy^3$$

$$\therefore \frac{\partial^2 \phi}{\partial y^2} = -\left(\frac{6Fx}{h^2}\right) + \left(\frac{12F}{h^3}\right)xz$$

and $\frac{\partial^2 \phi}{\partial x^2} = 0$

also $\frac{\partial^2 \phi}{\partial x \partial y} = -\left(\frac{6Fz}{h^2}\right) + \left(\frac{6F}{h^3}\right)y^2.$

Hence,

$$\sigma_x = -\left(\frac{6Fx}{h^2}\right) + \left(\frac{12F}{h^3}\right)xy \qquad (i)$$

$$\sigma_y = 0 \qquad (ii)$$

$$\tau_{xy} = -\frac{\partial^2 \phi}{\partial x \partial y} = -\left(\frac{6Fy}{h^2}\right) + \left(\frac{6F}{h^3}\right)y^2 \qquad (iii)$$

Variation of Stresses at Certain Boundary Points

(a) **Variation of σ_x**

From (i), it is clear that σ_x varies linearly with x, and at a given section it varies linearly with z.

$\therefore$ At $x = 0$ and $y = \pm h$, $\sigma_x = 0$

At $x = L$ and $y = 0$, $\sigma_x = -\left(\frac{6FL}{h^2}\right)$

At $x = L$ and $y = +h$, $\sigma_x = -\left(\frac{6FL}{h^2}\right) + \left(\frac{12F}{h^3}\right)Lh = \frac{6FL}{h^2}$

At $x = L$ and $y = -h$, $\sigma_x = -\left(\frac{6FL}{h^2}\right) - \left(\frac{12F}{h^3}\right)Lh = -\left(\frac{18FL}{h^2}\right)$

The variation of σ_x is shown in the Fig. 5.12.

(b) **Variation of σ_y**

σ_y is zero for all values of x.

(c) **Variation of τ_{xy}**

We have $\tau_{xy} = \left(\frac{6Fy}{h^2}\right) - \left(\frac{6F}{h^3}\right).y^2$

From the above expression, it is clear that the variation of τ_{xy} is parabolic with y. However, τ_{xy} is independent of x and is thus constant along the length, corresponding to a given value of y.

$\therefore$ At $y = 0$, $\tau_{xy} = 0$

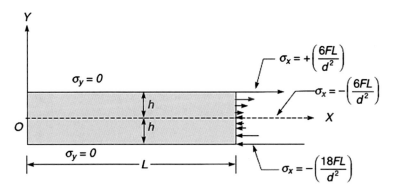

Fig. 5.12 Variation of σ_x

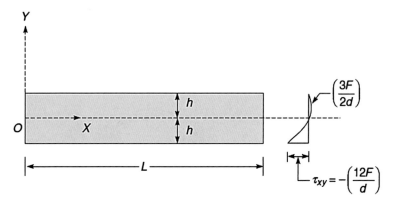

Fig. 5.13 Variation of τ_{xy}

At $y = + h$, $\tau_{xy} = \left(\frac{6Fh}{h^2}\right) - \left(\frac{6F}{h^3}\right)h^2 = 0$

At $y = y$, $\tau_{xy} = -\left(\frac{6F}{h^2}\right)h - \left(\frac{6F}{h^3}\right)(-h^2) = -\left(\frac{12F}{h}\right)$

The variation of τ_{xy} is shown in Fig. 5.13.

Example 5.4

Investigate what problem of plane stress is satisfied by the stress function

$$\phi = \frac{3F}{4h}\left[xy - \frac{xy^3}{3h^2}\right] + \frac{p}{2}y^2$$

applied to the region included in $y = 0$, $y = h$, $x = 0$ on the side x positive.

Solution The given stress function may be written as

$$\phi = \left(\frac{3F}{4h}\right)xy - \left(\frac{1}{4}\frac{Fxy^3}{h^3}\right) + \left(\frac{p}{2}\right)y^2$$

Now

$$\frac{\partial^2 \phi}{\partial x^2} = 0$$

$$\frac{\partial^2 \phi}{\partial y^2} = -\left(\frac{3 \times 2}{4} \cdot \frac{Fxy}{h^3}\right) + \frac{2p}{2} = p - \left(1.5\frac{F}{h^3}\right)xy$$

and

$$\frac{\partial^2 \phi}{\partial x \partial y} = \frac{3F}{4h} - \frac{3}{4}\frac{Fy^2}{h^3}$$

Hence, the stress components are

$$\sigma_x = \frac{\partial^2 \phi}{\partial y^2} = p - 1.5\frac{F}{h^3}xy$$

$$\sigma_y = \frac{\partial^2 \phi}{\partial x^2} = 0$$

$$\tau_{xy} = -\frac{\partial^2 \phi}{\partial x \partial y} = \frac{3}{4}\frac{Fy^2}{h^3} - \frac{3F}{4h}$$

(a) **Variation of σ_x (Fig 5.14)**

$$\sigma_x = p - \left(1.5\frac{F}{h^3}\right)xy$$

When $x = 0$ and $y = 0$ or $\pm h$, $\sigma_x = p$ (i.e. constant across the section)
When $x = L$ and $y = 0$, $\sigma_x = p$
When $x = L$ and $y = +h$, $\sigma_x = p - \left(1.5\frac{FL}{h^2}\right)$
When $x = L$ and $y = -h$, $\sigma_x = p + 1.5\frac{FL}{h^2}$
Thus, at $x = L$, the variation of σ_x is linear with y.
The variation of σ_x is shown in the figure below.

(b) **Variation of σ_y**

$$\sigma_y = \frac{\partial^2 \phi}{\partial x^2} = 0$$

$\therefore \sigma_y$ is zero for all value of x and y

(c) **Variation of τ_{xy}**

$$\tau_{xz} = \left(\frac{3}{4}\frac{Fy^2}{h^3}\right) - \left(\frac{3}{4}\frac{F}{h}\right)$$

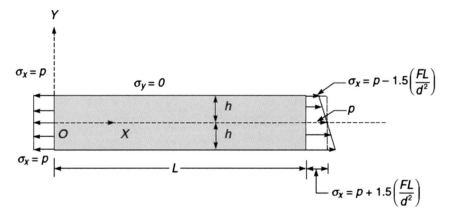

Fig. 5.14 Variation of stress σ_x

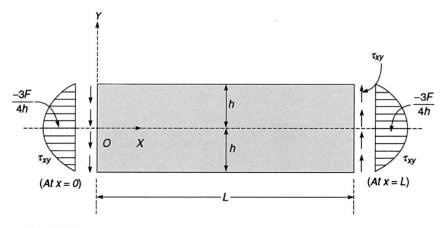

Fig. 5.15 Variation of shear stress τ_{xy}

Thus, τ_{xy} varies parabolically with z. However, it is independent of x; i.e. its value is the same for all values of x (Fig. 5.15).

$$\therefore \text{At} \quad y = 0, \quad \tau_{xy} = -\left(\frac{3}{4}\frac{F}{h}\right)$$

$$\text{At} \quad y = \pm h, \quad \tau_{xy} = \left[\frac{3}{4}\frac{F}{h^3}(h)^2\right] - \left[\frac{3}{4}\frac{F}{h}\right] = 0$$

5.10   Exercises

1. Write short notes on the following.

   (i)   Plane stress and plain-strain problems.
   (ii) Airy's stress function

2. Explain the use of polynomials in the solution of structural problems.
3. Determine the stress fields that arise from the following stress functions

   (a) $\phi = cy^2$ (b) $\phi = Ax^2 + Bxy + Cy^2$ (c) $\phi = Ax^3 + Bx^2y + Cxy^2 + Dy^3$

4. Investigate what problem of plane stress is solved by the following stress function.

$$\phi = \frac{F}{d^3}xy^2(3d - 2y)$$

5. Discuss what problems of plane stress can be solved using at a 4th degree polynomial.

6. Using stress function method, obtain the expressions for the stresses in a cantilever beam fixed $x = 0$; and carrying a concentrated load P at its free end. The beam is of rectangular cross section of width b and depth d.

7. Derive the compatibility equation n terms of stress components for plane stress problems when body forces are not constant.

8. Check whether the following is a stress function.

$$\phi = \frac{3}{4}\left\{ xy - \frac{xy^2}{c} - \frac{xy^3}{c^2} + \frac{ly^2}{c} + \frac{ly^3}{c^2} \right\}$$

9. Given the following polynomial:

$$\phi = C_1 x^4 + C_2 x^3 y + C_3 x^2 y^2 + C_4 xy^3 + C_5 y^4$$

adjust the coefficient so that the function is a Biharmonic function.
Now adjust the constants so that there is a solution for the cantilever beam shown in Fig. 5.16, where uniform shear stress tractions are applied at the upper and lower edge of the beam and a point load P is applied at the tip. Explain the limitations as to domain of problems for which the solution is valid within the plane stress theory.

10. Investigate what kind of problem is solved the stress function

$$\phi = -\left(\frac{w}{d^3}\right)xy^2(3d - 2y)\text{to the region } 0 \le y \le d, \quad 0 \le x$$

11. Determine the elasticity problem that is solved by the stress function

$$\phi = Ax^3 y \quad \text{for the region } -a \le x \le a, \quad -b \le y \le b$$

12. For the rectangle shown in Fig. 5.17, the following stress functions are considered.

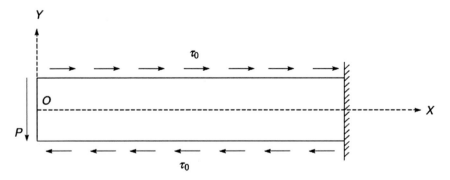

Fig. 5.16 Beam subjected to uniform shear stress

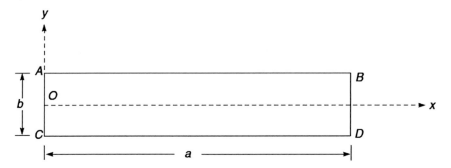

Fig. 5.17 Rectangular beam

(i) $\phi = C_1 x^2$
(ii) $\phi = C_1 xy$
(iii) $\phi = C_1 y^3$
(iv) $\phi = \frac{C_1}{2}x^2 + C_2 xy + \frac{C_3}{2}y^2$

Investigate for the stress field and comment.

13. A thin space plate whose sides are parallel to x and y axes the following distri-
     bution $\sigma_x = Ay$; $\sigma_y = Ax$ and some shear stress where A is a constant. Find the
     suitable stress function and the nature of shear stress which can be associated
     with the given normal stresses.

Chapter 6
Two-Dimensional Problems in Elasticity (in Polar Coordinate System)

6.1 Introduction

In any elasticity problem, the proper choice of the co-ordinate system is extremely important since this choice establishes the complexity of the mathematical expressions employed to satisfy the field equations and the boundary conditions.

In order to solve two-dimensional elasticity problems by employing a polar co-ordinate reference frame, the equations of equilibrium, the definition of Airy's stress function and one of the stress equations of compatibility must be established in terms of polar co-ordinates.

6.2 Strain–Displacement Relations

Case 1: For Two-Dimensional State of Stress.

Consider the deformation of the infinitesimal element $ABCD$, denoting r and θ displacements by u and v respectively. The general deformation experienced by an element may be regarded as composed of (1) a change in the length of the sides, and (2) rotation of the sides as shown in Fig. 6.1.

Referring to the figure, it is observed that a displacement "u" of side AB results in both radial and tangential strains.

Therefore,

$$\text{radial strain} = \varepsilon_r = \frac{\partial u}{\partial r} \tag{6.1}$$

and tangential strain due to u per unit length of AB is

$$(\varepsilon_\theta)_u = \frac{(r + u)\mathrm{d}\theta - r\,\mathrm{d}\theta}{r\,\mathrm{d}\theta} = \frac{u}{r} \tag{6.2}$$

Fig. 6.1 Deformed element
in two dimensions

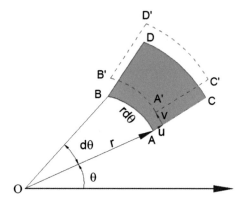

Tangential strain due to displacement v is given by

$$(\varepsilon_\theta)_v = \frac{\left(\frac{\partial v}{\partial \theta}\right)d\theta}{rd\theta} = \frac{1}{r}\frac{\partial v}{\partial \theta} \tag{6.3}$$

Hence, the resultant strain is

$$\varepsilon_\theta = (\varepsilon_\theta)u + (\varepsilon_\theta)_v$$

$$\varepsilon_\theta = \frac{u}{r} + \frac{1}{r}\left(\frac{\partial v}{\partial \theta}\right) \tag{6.4}$$

Similarly, the shearing strains can be calculated due to displacements u and v as below.

Component of shearing strain due to u is

$$(\gamma_{r\theta})_u = \frac{\left(\frac{\partial u}{\partial \theta}\right)d\theta}{rd\theta} = \frac{1}{r}\left(\frac{\partial u}{\partial \theta}\right) \tag{6.5}$$

Component of shearing strain due to v is

$$(\gamma_{r\theta})_v = \frac{\partial v}{\partial r} - \left(\frac{v}{r}\right) \tag{6.6}$$

Therefore, the total shear strain is given by

$$\gamma_{r\theta} = (\gamma_{r\theta})_u + (\gamma_{r\theta})_v$$

$$\gamma_{r\theta} = \frac{1}{r}\left(\frac{\partial u}{\partial \theta}\right) + \frac{\partial v}{\partial r} - \left(\frac{v}{r}\right) \tag{6.7}$$

Case 2: For Three-Dimensional State of Stress.
See Fig. 6.2.

Fig. 6.2 Deformed element
in three dimensions

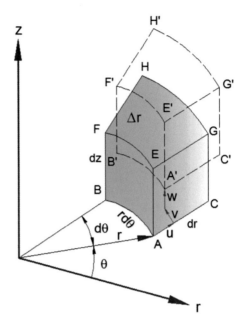

The strain–displacement relations for the most general state of stress are given by

$$\varepsilon_r = \frac{\partial u}{\partial r}, \quad \varepsilon_\theta = \frac{1}{r}\left(\frac{\partial v}{\partial \theta}\right) + \left(\frac{u}{r}\right), \quad \varepsilon_z = \frac{\partial w}{\partial z}$$

$$\gamma_{r\theta} = \frac{\partial v}{\partial r} + \frac{1}{r}\left(\frac{\partial u}{\partial \theta}\right) - \left(\frac{v}{r}\right)$$

$$\gamma_{\theta_z} = \frac{1}{r}\left(\frac{\partial w}{\partial \theta}\right) + \left(\frac{\partial v}{\partial z}\right)$$

$$\gamma_{zr} = \frac{\partial u}{\partial z} + \left(\frac{\partial w}{\partial r}\right) \tag{6.8}$$

6.3 Strain-Compatibility Equation

We have from the strain–displacement relations:

$$\text{Radial strain, } \varepsilon_r = \frac{\partial u}{\partial r} \tag{6.9a}$$

$$\text{Tangential strain, } \varepsilon_\theta = \left(\frac{1}{r}\right)\frac{\partial v}{\partial \theta} + \left(\frac{u}{r}\right) \tag{6.9b}$$

and

$$\text{total shearing strain, } \gamma_{r\theta} = \frac{\partial v}{\partial r} - \left(\frac{v}{r}\right) + \left(\frac{1}{r}\right)\frac{\partial u}{\partial \theta} \tag{6.9c}$$

Differentiating Eq. (6.9a) with respect to θ and Eq. (6.9b) with respect to r, we get

$$\frac{\partial \varepsilon_r}{\partial \theta} = \frac{\partial^2 u}{\partial r \partial \theta} \tag{6.9d}$$

$$\frac{\partial \varepsilon_\theta}{\partial r} = \left(\frac{1}{r}\right)\frac{\partial u}{\partial r} - \left(\frac{1}{r^2}\right)u + \frac{1}{r}\cdot\frac{\partial^2 v}{\partial r \partial \theta} - \left(\frac{1}{r^2}\right)\cdot\frac{\partial v}{\partial \theta}$$

$$= \frac{\varepsilon_r}{r} + \left(\frac{1}{r}\right)\frac{\partial^2 v}{\partial r \partial \theta} - \frac{1}{r}\left[\frac{u}{r} + \left(\frac{1}{r}\right)\frac{\partial v}{\partial \theta}\right]$$

$$\therefore \frac{\partial \varepsilon_\theta}{\partial r} = \frac{\varepsilon_r}{r} + \left(\frac{1}{r}\right)\cdot\frac{\partial^2 v}{\partial r \partial \theta} - \left(\frac{1}{r}\right)\varepsilon_\theta \tag{6.9e}$$

Now, differentiating Eq. (6.9c) with respect to r and using Eq. (6.9d), we get

$$\frac{\partial \gamma_{r\theta}}{\partial r} = \frac{\partial^2 v}{\partial r^2} - \left(\frac{1}{r}\right)\frac{\partial v}{\partial r} + \frac{v}{r^2} + \left(\frac{1}{r}\right)\frac{\partial^2 u}{\partial r \partial \theta} - \left(\frac{1}{r^2}\right)\frac{\partial u}{\partial \theta}$$

$$= \frac{\partial^2 v}{\partial r^2} - \frac{1}{r}\left(\frac{\partial v}{\partial r} - \frac{v}{r} + \frac{1}{r}\frac{\partial u}{\partial \theta}\right) + \frac{1}{r}\frac{\partial^2 u}{\partial r \partial \theta}$$

$$\therefore \frac{\partial \gamma_{r\theta}}{\partial r} = \frac{\partial^2 v}{\partial r^2} - \left(\frac{1}{r}\right)\gamma_{r\theta} + \left(\frac{1}{r}\right)\frac{\partial \varepsilon_r}{\partial \theta} \tag{6.9f}$$

Differentiating Eq. (6.9e) with respect to r and Eq. (6.9f) with respect to θ, we get

$$\frac{\partial^2 \varepsilon_\theta}{\partial r^2} = \left(\frac{1}{r}\right)\frac{\partial \varepsilon_r}{\partial r} - \left(\frac{1}{r^2}\right)\varepsilon_r + \left(\frac{1}{r}\right)\frac{\partial^3 v}{\partial r^2 \partial \theta} - \left(\frac{1}{r^2}\right)\frac{\partial^2 v}{\partial r \partial \theta} - -\left(\frac{1}{r}\right)\frac{\partial \varepsilon_\theta}{\partial r} + \frac{1}{r^2}\varepsilon_\theta \tag{6.9g}$$

and

$$\frac{\partial^2 \gamma_{r\theta}}{\partial r \partial \theta} = \frac{\partial^3 v}{\partial r^2 \partial \theta} - \left(\frac{1}{r}\right)\frac{\partial \gamma_{r\theta}}{\partial \theta} + \left(\frac{1}{r}\right)\frac{\partial^2 \varepsilon_r}{\partial \theta^2}$$

or

$$\left(\frac{1}{r}\right)\frac{\partial^2 \gamma_{r\theta}}{\partial r \partial \theta} = \left(\frac{1}{r}\right)\frac{\partial^3 v}{\partial r^2 \partial \theta} - \left(\frac{1}{r^2}\right)\frac{\partial \gamma_{r\theta}}{\partial \theta} + \left(\frac{1}{r^2}\right)\frac{\partial^2 \varepsilon_r}{\partial \theta^2} \tag{6.9h}$$

Subtracting Eq. (6.9h) from Eq. (6.9g) and using Eq. (6.9e), we get

$$
\frac{\partial^2 \varepsilon_\theta}{\partial r^2} - \left(\frac{1}{r}\right)\frac{\partial^2 \gamma_{r\theta}}{\partial r \partial \theta} = \left(\frac{1}{r}\right)\frac{\partial \varepsilon_r}{\partial r} - \left(\frac{\varepsilon_r}{r^2}\right) - \left(\frac{1}{r^2}\right)\frac{\partial^2 v}{\partial r \partial \theta} - \left(\frac{1}{r}\right)\frac{\partial \varepsilon_\theta}{\partial r}
$$

$$
+ \left(\frac{1}{r^2}\right)\frac{\partial \gamma_{r\theta}}{\partial \theta} - \left(\frac{1}{r^2}\right)\frac{\partial^2 \varepsilon_r}{\partial \theta^2} + \frac{\varepsilon_\theta}{r^2}
$$

$$
= \frac{1}{r}\left(\frac{\partial \varepsilon_r}{\partial r}\right) - \frac{1}{r}\left(\frac{\varepsilon_r}{r} + \frac{1}{r}\frac{\partial^2 v}{\partial r \partial \theta} - \frac{\varepsilon_\theta}{r}\right)
$$

$$
- \frac{1}{r}\left(\frac{\partial \varepsilon_\theta}{\partial r} - \frac{1}{r}\cdot\frac{\partial \gamma_{r\theta}}{\partial \theta} + \frac{1}{r}\frac{\partial^2 \varepsilon_r}{\partial \theta^2}\right)
$$

$$
= \left(\frac{1}{r}\right)\frac{\partial \varepsilon_r}{\partial r} - \left(\frac{1}{r}\right)\frac{\partial \varepsilon_\theta}{\partial r} - \left(\frac{1}{r}\right)\frac{\partial \varepsilon_\theta}{\partial r}
$$

$$
+ \left(\frac{1}{r^2}\right)\frac{\partial \gamma_{r\theta}}{\partial \theta} - \left(\frac{1}{r^2}\right)\frac{\partial^2 \varepsilon_r}{\partial \theta^2}
$$

$$
= \left(\frac{1}{r}\right)\frac{\partial \varepsilon_r}{\partial r} - \left(\frac{2}{r}\right)\frac{\partial \varepsilon_\theta}{\partial r} + \left(\frac{1}{r^2}\right)\frac{\partial \gamma_{r\theta}}{\partial \theta} - \left(\frac{1}{r^2}\right)\frac{\partial^2 \varepsilon_r}{\partial \theta^2}
$$

$$
\therefore \left(\frac{1}{r^2}\right)\frac{\partial \gamma_{r\theta}}{\partial \theta} + \left(\frac{1}{r}\right)\frac{\partial^2 \gamma_{r\theta}}{\partial r \partial \theta} = \frac{\partial^2 \varepsilon_\theta}{\partial r^2} + \left(\frac{2}{r}\right)\frac{\partial \varepsilon_\theta}{\partial r} - \left(\frac{1}{r}\right)\frac{\partial \varepsilon_r}{\partial r} + \left(\frac{1}{r^2}\right)\frac{\partial^2 \varepsilon_r}{\partial \theta^2}
$$

6.4 Stress–Strain Relations

In terms of cylindrical coordinates, the stress–strain relations for three-dimensional state of stress and strain are given by

$$
\varepsilon_r = \frac{1}{E}[\sigma_r - v(\sigma_\theta + \sigma_z)]
$$

$$
\varepsilon_\theta = \frac{1}{E}[\sigma_\theta - v(\sigma_r + \sigma_z)]
$$

$$
\varepsilon_z = \frac{1}{E}[\sigma_z - v(\sigma_r + \sigma_\theta)] \tag{6.10}
$$

For two-dimensional state of stresses and strains, the above equations reduce to,
For Plane Stress Case

$$
\varepsilon_r = \frac{1}{E}(\sigma_r - v\sigma_\theta)
$$

$$
\varepsilon_\theta = \frac{1}{E}(\sigma_\theta - v\sigma_r)
$$

$$
\gamma_{r\theta} = \frac{1}{G}\tau_{r\theta} \tag{6.11}
$$

For Plane Strain Case

$$\varepsilon_r = \frac{(1+v)}{E}[(1-v)\sigma_r - v\sigma_\theta]$$

$$\varepsilon_\theta = \frac{(1+v)}{E}[(1-v)\sigma_\theta - v\sigma_r]$$

$$\gamma_r\theta = \frac{1}{G}\tau_{r\theta} \tag{6.12}$$

6.5   Airy's Stress Function

With reference to the two-dimensional equations or stress transformation [Eqs. (2.13a)–(2.13c)], the relationship between the polar stress components σ_r, σ_θ and $\tau_{r\theta}$ and the Cartesian stress components σ_x, σ_y and τ_{xy} can be obtained as below.

$$\sigma_r = \sigma_x \cos^2\theta + \sigma_y \sin^2\theta + \tau_{xy}\sin 2\theta$$

$$\sigma_\theta = \sigma_y \cos^2\theta + \sigma_x \sin^2\theta - \sin 2\theta$$

$$\tau_{r\theta} = (\sigma_y - \sigma_x)\sin\theta\cos\theta + \tau_{xy}\cos 2\theta \tag{6.13}$$

Now we have,

$$\sigma_x = \frac{\partial^2\phi}{\partial y^2} \quad \sigma_y = \frac{\partial^2\phi}{\partial x^2} \quad \tau_{xy} = -\frac{\partial^2\phi}{\partial x\partial y} \tag{6.14}$$

Substituting (6.14) in (6.13), we get

$$\sigma_r = \frac{\partial^2\phi}{\partial y^2}\cos^2\theta + \frac{\partial^2\phi}{\partial x^2}\sin^2\theta - \frac{\partial^2\phi}{\partial x\partial y}\sin 2\theta$$

$$\sigma_\theta = \frac{\partial^2\phi}{\partial x^2}\cos^2\theta + \frac{\partial^2\phi}{\partial y^2}\sin^2\theta + \frac{\partial^2\phi}{\partial x\partial y}\sin 2\theta$$

$$\tau_{r\theta} = \left(\frac{\partial^2\phi}{\partial x^2} - \frac{\partial^2\phi}{\partial y^2}\right)\sin\theta\cos\theta - \frac{\partial^2\phi}{\partial x\partial y}\cos 2\theta \tag{6.15}$$

The polar components of stress in terms of Airy's stress functions are as follows.

$$\sigma_r = \left(\frac{1}{r}\right)\frac{\partial\phi}{\partial r} + \left(\frac{1}{r^2}\right)\frac{\partial^2\phi}{\partial\theta^2} \tag{6.16}$$

$$\sigma_\theta = \frac{\partial^2\phi}{\partial r^2} \quad \text{and} \quad \tau_{r\theta} = \left(\frac{1}{r^2}\right)\frac{\partial\phi}{\partial\theta} - \left(\frac{1}{r}\right)\frac{\partial^2\phi}{\partial r\partial\theta} \tag{6.17}$$

The above relations can be employed to determine the stress field as a function of r and θ.

6.6 Biharmonic Equation

As discussed earlier, the Airy's stress function ϕ has to satisfy the biharmonic equation $\nabla^4\phi = 0$, provided the body forces are zero or constants. In polar coordinates, the stress function must satisfy this same equation; however, the definition of ∇^4 operator must be modified to suit the polar co-ordinate system. This modification may be accomplished by transforming the ∇^4 operator from the Cartesian system to the polar system.

Now, we have, $x = r\cos\theta, \quad y = r\sin\theta$

$$r^2 = x^2 + y^2 \text{ and } \theta = \tan^{-1}\left(\frac{y}{x}\right) \tag{6.18}$$

where r and θ are defined in Fig. 6.3.
Differentiating Eq. (6.18) gives

$$\frac{\partial r}{\partial x} = \frac{x}{r} = \frac{r\cos\theta}{r} = \cos\theta$$

$$\frac{\partial r}{\partial y} = \frac{y}{r} = \frac{r\sin\theta}{r} = \sin\theta$$

$$\frac{\partial\theta}{\partial x} = -\left(\frac{y}{r^2}\right) = \frac{r\sin\theta}{r^2} = -\left(\frac{\sin\theta}{r}\right)$$

$$\frac{\partial\theta}{\partial y} = \frac{x}{r^2} = \frac{r\cos\theta}{r^2} = \frac{\cos\theta}{r}$$

Fig. 6.3 Generalized co-ordinate system

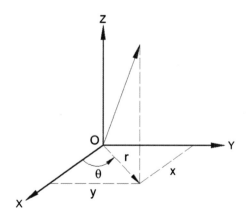

$$2\,r\,\mathrm{d}r = 2x\mathrm{d}x + 2y\,\mathrm{d}y$$

$$\therefore \mathrm{d}r = \left(\frac{x}{r}\right)\mathrm{d}x + \left(\frac{y}{r}\right)\mathrm{d}y$$

Also,

$$\sec^2\theta\ \mathrm{d}\theta = -\left(\frac{y}{x^2}\right)xy + \left(\frac{\mathrm{d}y}{x}\right)$$

$$\frac{\partial\phi}{\partial x} = \frac{\partial\phi}{\partial r}\frac{\partial r}{\partial x} + \frac{\partial\phi}{\partial\theta}\frac{\partial\theta}{\partial x}$$

$$= \frac{x}{\sqrt{x^2+y^2}}\frac{\partial\phi}{\partial r} + \left(-\frac{1}{\sec^2\theta}\left(\frac{y}{x^2}\right)\right)\frac{\partial\phi}{\partial\theta}$$

$$\therefore \frac{\partial\phi}{\partial x} = \cos\theta\left(\frac{\partial\phi}{\partial r}\right) - \frac{\sin\theta}{r}\left(\frac{\partial\phi}{\partial\theta}\right)$$

Similarly,

$$\frac{\partial\phi}{\partial y} = \frac{\partial\phi}{\partial r}\frac{\partial r}{\partial y} + \frac{\partial\phi}{\partial\theta}\frac{\partial\theta}{\partial y}$$

$$\therefore \frac{\partial\phi}{\partial y} = \sin\theta\left(\frac{\partial\phi}{\partial r}\right) + \frac{\cos\theta}{r}\left(\frac{\partial\phi}{\partial\theta}\right)$$

Now,

$$\frac{\partial^2\phi}{\partial x^2} = \left(\cos\theta\left(\frac{\partial\phi}{\partial r}\right) - \frac{\sin\theta}{r}\left(\frac{\partial\phi}{\partial\theta}\right)\right)^2$$

$$= \cos^2\theta\frac{\partial^2\phi}{\partial r^2} - \left(\frac{2\sin\theta\cos\theta}{r}\right)\frac{\partial^2\phi}{\partial r\partial\theta} + \frac{\sin^2\theta}{r^2}\left(\frac{\partial^2\phi}{\partial\theta^2}\right)$$

$$+ \frac{2\sin\theta\cos\theta}{r^2}\left(\frac{\partial\phi}{\partial\theta}\right) + \frac{\sin^2\theta}{r}\left(\frac{\partial\phi}{\partial r}\right) \tag{i}$$

Similarly,

$$\frac{\partial^2\phi}{\partial y^2} = \sin^2\theta\frac{\partial^2\phi}{\partial r^2} + \frac{2\sin\theta\cos\theta}{r}\frac{\partial^2\phi}{\partial r\partial\theta} - \left(\frac{2\sin\theta\cos\theta}{r^2}\right)\frac{\partial\phi}{\partial\theta}$$

$$+ \frac{\cos^2\theta}{r}\left(\frac{\partial\phi}{\partial r}\right) + \frac{\cos^2\theta}{r^2}\frac{\partial^2\phi}{\partial\theta^2} \tag{ii}$$

And,

$$\frac{\partial^2\phi}{\partial x\partial y} = -\left(\frac{\sin\theta\cos\theta}{r}\right)\frac{\partial\phi}{\partial r} + \sin\theta\cos\theta\frac{\partial^2\phi}{\partial r^2} + \frac{\cos 2\theta}{r}\frac{\partial^2\phi}{\partial r\partial\theta}$$

$$-\left(\frac{\cos 2\phi}{r^2}\right)\frac{\partial\phi}{\partial\theta} - \left(\frac{\sin\theta\cos\theta}{r^2}\right)\frac{\partial^2\phi}{\partial\theta^2} \tag{iii}$$

Adding (i) and (ii), we get

$$\frac{\partial^2\phi}{\partial x^2} + \frac{\partial^2\phi}{\partial y^2} = \frac{\partial^2\phi}{\partial r^2} + \left(\frac{1}{r}\right)\frac{\partial\phi}{\partial r} + \left(\frac{1}{r^2}\right)\frac{\partial^2\phi}{\partial\theta^2}$$

i.e., $\nabla^2\phi = \dfrac{\partial^2\phi}{\partial x^2} + \dfrac{\partial^2\phi}{\partial y^2} = \dfrac{\partial^2\phi}{\partial r^2} + \left(\dfrac{1}{r}\right)\dfrac{\partial\phi}{\partial r} + \left(\dfrac{1}{r^2}\right)\dfrac{\partial^2\phi}{\partial\theta^2}$

or $\nabla^4\phi = \nabla^2\phi\nabla^2\phi = \left(\dfrac{\partial^2\phi}{\partial r^2} + \dfrac{1}{r}\dfrac{\partial\phi}{\partial r} + \dfrac{1}{r^2}\dfrac{\partial^2\phi}{\partial\theta^2}\right)\left(\dfrac{\partial^2\phi}{\partial r^2} + \dfrac{1}{r}\dfrac{\partial\phi}{\partial r} + \dfrac{1}{r^2}\dfrac{\partial^2\phi}{\partial\theta^2}\right) = 0$

The above biharmonic equation is the stress equation of compatibility in terms of Airy's stress function referred to a polar co-ordinate system.

6.7 Axisymmetric Problems

Many engineering problems involve solids of revolution subjected to axially symmetric loading. The examples are a circular cylinder loaded by uniform internal or external pressure or other axially symmetric loading (Fig. 6.4a), and a semi-infinite half space loaded by a circular area, for example, a circular footing on a soil mass (Fig. 6.4b). It is convenient to express these problems in terms of the cylindrical co-ordinates. Because of symmetry, the stress components are independent of the angular (θ) co-ordinate; hence, all derivatives with respect to θ vanish and the components v, $\gamma_{r\theta}$, $\gamma_{\theta z}$, $\tau_{r\theta}$ and $\tau_{\theta z}$ are zero. The nonzero stress components are σ_r, σ_θ, σ_z and τ_{rz}.

The strain–displacement relations for the nonzero strains become

$$\varepsilon_r = \frac{\partial u}{\partial r}, \varepsilon_\theta = \frac{u}{r}, \varepsilon_z = \frac{\partial w}{\partial z}$$

$$\gamma_{rz} = \frac{\partial u}{\partial z} + \frac{\partial w}{\partial r} \tag{6.19}$$

and the constitutive relation is given by

$$\begin{Bmatrix} \sigma_r \\ \sigma_z \\ \sigma_\theta \\ \tau_{rz} \end{Bmatrix} = \frac{E}{(1+v)(1-2v)} \begin{bmatrix} (1-v) & v & v & 0 \\ & (1-v) & v & 0 \\ & & (1-v) & 0 \\ & \text{Symmetry} & & \frac{(1-2v)}{2} \end{bmatrix} \begin{Bmatrix} \varepsilon_r \\ \varepsilon_z \\ \varepsilon_\theta \\ \gamma_{rz} \end{Bmatrix}$$

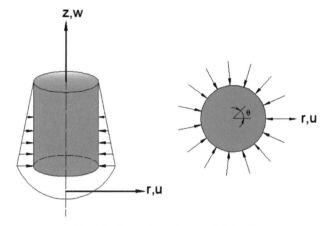

(a) Cylinder under axisymmetric loading

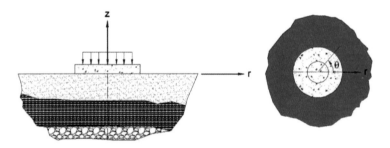

(b) Circular footing on soil mass

Fig. 6.4 Axisymmetric problems

6.8 Thick-Walled Cylinder Subjected to Internal and External Pressures

Consider a cylinder of inner radius "a" and outer radius "b" as shown in Fig. 6.5. Let the cylinder be subjected to internal pressure p_i and an external pressure p_0. This problem can be treated either as a plane stress case ($\sigma_z = 0$) or as a plane strain case ($\varepsilon_z = 0$).

Case (a): Plane stress.

Consider the ends of the cylinder which are free to expand. As the axial loading is absent, $\sigma_z = 0$. The stresses are symmetrical about the z-axis and therefore, $\tau_{r\theta} = 0$. Neglecting the body forces, Eq. (2.46) reduces to

$$\frac{\partial \sigma_r}{\partial r} + \left(\frac{\sigma_r - \sigma_\theta}{r} \right) = 0 \tag{6.20}$$

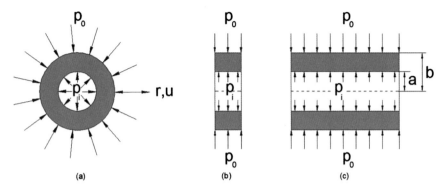

Fig. 6.5 a Thick-walled cylinder, **b** plane stress case and **c** plane strain case

Here σ_θ and σ_r denote tangential (circumferential) and radial stresses acting normal to the sides of the element.

Since r is the only independent variable, the above equation can be written as

$$\frac{d}{dr}(r\sigma_r) - \sigma_\theta = 0. \tag{6.21}$$

From Hooke's law,

$$\varepsilon_r = \frac{1}{E}(\sigma_r - \nu\sigma_\theta), \quad \varepsilon_\theta = \frac{1}{E}(\sigma_\theta - \nu\sigma_r)$$

Also,

$$\varepsilon_r = \frac{du}{dr} \text{ and } \varepsilon_\theta = \frac{u}{r}$$

Hence, the stresses in terms of strains are

$$\sigma_r = \frac{E}{(1-\nu^2)}(\varepsilon_r + \nu\varepsilon_\theta)$$

$$\sigma_\theta = \frac{E}{(1-\nu^2)}(\varepsilon_\theta + \nu\varepsilon_r)$$

Substituting the values of ε_r and ε_θ in the above expressions, we get

$$\sigma_r = \frac{E}{(1-\nu^2)}\left(\frac{du}{dr} + \nu\frac{u}{r}\right)$$

$$\sigma_\theta = \frac{E}{(1-\nu^2)}\left(\frac{u}{r} + \nu\frac{du}{dr}\right)$$

Substituting these in the equilibrium Eq. (6.21), then

$$\frac{d}{dr}\left(r\frac{du}{dr} + vu\right) - \left(\frac{u}{r} + v\frac{du}{dr}\right) = 0$$

$$\frac{du}{dr} + r\frac{d^2u}{dr^2} + v\frac{du}{dr} - \frac{u}{r} - v\frac{du}{dr} = 0$$

$$\text{or } \frac{d^2u}{dr^2} + \frac{1}{r}\frac{du}{dr} - \frac{u}{r^2} = 0$$

The above equation is called equidimensional equation in radial displacement. The solution of the above equation is

$$U = C_1 r + C_2/r \tag{6.22}$$

where C_1 and C_2 are constants.

The radial and tangential stresses are written in terms of constants of integration C_1 and C_2.

Therefore,

$$\sigma_r = \frac{E}{(1 - v^2)}\left[C_1(1 + v) - C_2\left(\frac{1 - v}{r^2}\right)\right]$$

$$\sigma_\theta = \frac{E}{(1 - v^2)}\left[C_1(1 + v) + C_2\left(\frac{1 - v}{r^2}\right)\right] \tag{6.23}$$

The constants are determined from the boundary conditions.

when

$$\begin{aligned} r = a, \sigma_r = -p_i \\ r = b, \sigma_r = -p_0 \end{aligned} \tag{6.23a}$$

Hence,

$$\frac{E}{(1 - v^2)}\left[C_1(1 + v) - C_2\left(\frac{1 - v}{a^2}\right)\right] = -P_i$$

$$\text{and } \frac{E}{(1 - v^2)}\left[C_1(1 + v) - C_2\left(\frac{1 - v}{b^2}\right)\right] = -P_0$$

where the negative sign in the boundary conditions denotes compressive stress. The constants are evaluated by substitution of Eq. (6.23a) into (6.23)

$$C_1 = \left(\frac{1 - v}{E}\right)\left(\frac{a^2 p_i - b^2 p_0}{(b^2 - a^2)}\right)$$

$$C_2 = \left(\frac{1 + v}{E}\right)\left(\frac{a^2 b^2(p_i - p_0)}{(b^2 - a^2)}\right)$$

Substituting these in Eqs. (6.22) and (6.23), we get

$$\sigma_r = \left(\frac{a^2 p_i - b^2 p_0}{b^2 - a^2}\right) - \left(\frac{(p_i - p_0)a^2b^2}{(b^2 - a^2)r^2}\right) \tag{6.24}$$

$$\sigma_\theta = \left(\frac{a^2 p_i - b^2 p_0}{b^2 - a^2}\right) + \left(\frac{(p_i - p_0)a^2b^2}{(b^2 - a^2)r^2}\right) \tag{6.25}$$

$$u = \left(\frac{1 - v}{E}\right)\frac{(a^2 p_i - b^2 p_0)r}{(b^2 - a^2)} + \left(\frac{1 + v}{E}\right)\frac{(p_i - p_0)a^2b^2}{(b^2 - a^2)r} \tag{6.26}$$

These expressions were first derived by G. Lambe.

It is interesting to observe that the sum $(\sigma_r + \sigma_\theta)$ is constant through the thickness of the wall of the cylinder, regardless of radial position. Hence according to Hooke's law, the stresses σ_r and σ_θ produce a uniform extension or contraction in z-direction. The cross-sections perpendicular to the axis of the cylinder remain plane. If two adjacent cross-sections are considered, then the deformation undergone by the element does not interfere with the deformation of the neighbouring element. Hence, the elements are considered to be in the plane stress state.

Special Cases.

(i) *A cylinder subjected to internal pressure only:* In this case, $p_0 = 0$ and $p_i = p$.

Then Eqs. (6.24) and (6.25) become

$$\sigma_r = \frac{pa^2}{(b^2 - a^2)}\left(1 - \frac{b^2}{r^2}\right) \tag{6.27}$$

$$\sigma_\theta = \frac{pa^2}{(b^2 - a^2)}\left(1 + \frac{b^2}{r^2}\right) \tag{6.28}$$

Fig. 6.6 Cylinder subjected to internal pressure

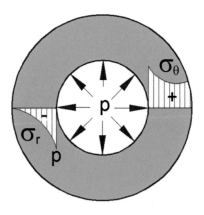

Figure 6.6 shows the variation of radial and circumferential stresses across the thickness of the cylinder under internal pressure.

The circumferential stress is greatest at the inner surface of the cylinder and is given by

$$(\sigma_\theta)_{max} = \frac{p(a^2 + b^2)}{b^2 - a^2} \tag{6.29}$$

(ii) *A cylinder subjected to external pressure only:* In this case, $p_i = 0$ and $p_0 = p$.

Equation (6.25) becomes

$$\sigma_r = -\left(\frac{pb^2}{b^2 - a^2}\right)\left(1 - \frac{a^2}{r^2}\right) \tag{6.30}$$

$$\sigma_\theta = -\left(\frac{pb^2}{b^2 - a^2}\right)\left(1 + \frac{a^2}{r^2}\right) \tag{6.31}$$

Figure 6.7 represents the variation of σ_r and σ_θ across the thickness.

However, if there is no inner hole, i.e., if $a = 0$, the stresses are uniformly distributed in the cylinder as.

$$\sigma_x = \sigma_\theta = -p.$$

Case (b): Plane Strain.

If a long cylinder is considered, sections that are far from the ends are in a state of plane strain and hence σ_z does not vary along the z-axis.

Fig. 6.7 Cylinder subjected to external pressure

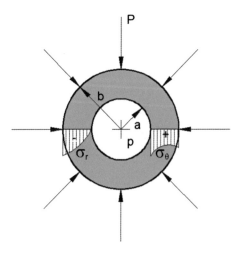

Now, from Hooke's law,

$$\varepsilon_r = \frac{1}{E}[\sigma_r - v(\sigma_\theta + \sigma_z)]$$

$$\varepsilon_\theta = \frac{1}{E}[\sigma_\theta - v(\sigma_r + \sigma_z)]$$

$$\varepsilon_z = \frac{1}{E}[\sigma_z - v(\sigma_r + \sigma_\theta)]$$

Since $\varepsilon_z = 0$, then

$$0 = \frac{1}{E}[\sigma_z - v(\sigma_r + \sigma_\theta)]$$

$$\sigma_{\bar{z}} = v(\sigma_r + \sigma_\theta)$$

Hence,

$$\varepsilon_r = \frac{(1+v)}{E}[(1-v)\sigma_r - v\sigma_\theta]$$

$$\varepsilon_\theta = \frac{(1+v)}{E}[(1-v)\sigma_\theta - v\sigma_r]$$

Solving for σ_θ and σ_r,

$$\sigma_\theta = \frac{E}{(1-2v)(1+v)}[v\varepsilon_r + (1-v)\varepsilon_\theta]$$

$$\sigma_r = \frac{E}{(1-2v)(1+v)}[(1-v)\varepsilon_r + v\varepsilon_\theta]$$

Substituting the values of ε_r and ε_θ, the above expressions for σ_θ and σ_r can be written as

$$\sigma_\theta = \frac{E}{(1-2v)(1+v)}\left[v\frac{du}{dr} + (1-v)\frac{u}{r}\right]$$

$$\sigma_r = \frac{E}{(1-2v)(1+v)}\left[(1-v)\frac{du}{dr} + \frac{vu}{r}\right]$$

Substituting these in the equation of equilibrium (6.21), we get

$$\frac{d}{dr}\left[(1-v)r\frac{du}{dr} + vu\right] - v\frac{du}{dr} - (1-v)\frac{u}{r} = 0$$

$$\text{or } \frac{du}{dr} + r\frac{d^2u}{dr^2} - \frac{u}{r} = 0$$

$$\frac{d^2u}{dr^2} + \frac{1}{r}\frac{du}{dr} - \frac{u}{r^2} = 0$$

The solution of this equation is the same as in Eq. (6.22)

$u = C_1 r + C_2/r.$

where C_1 and C_2 are constants of integration. Therefore, σ_θ and σ_r are given by

$$\sigma_\theta = \frac{E}{(1-2v)(1+v)}\left[C_1 + (1-2v)\frac{C_2}{r^2}\right]$$

$$\sigma_r = \frac{E}{(1-2v)(1+v)}\left[C_1 - (1-2v)\frac{C_2}{r^2}\right]$$

Applying the boundary conditions,

$$\sigma_r = -p_i \text{ when } r = a$$
$$\sigma_r = -p_0 \text{ when } r = b$$

Therefore,

$$\frac{E}{(1-2v)(1+v)}\left[C_1 - (1-2v)\frac{C_2}{a^2}\right] = -p_i$$

$$\frac{E}{(1-2v)(1+v)}\left[C_1 - (1-2v)\frac{C_2}{b^2}\right] = -p_o$$

Solving, we get

$$C_1 = \frac{(1-2v)(1+v)}{E}\left(\frac{p_0 b^2 - p_i a^2}{a^2 - b^2}\right)$$

and

$$C_2 = \frac{(1+v)}{E}\left(\frac{(p_0 - p_i)a^2 b^2}{a^2 - b^2}\right)$$

Substituting these, the stress components become

$$\sigma_r = \left(\frac{p_i a^2 - p_0 b^2}{b^2 - a^2}\right) - \left(\frac{p_i - p_0}{b^2 - a^2}\right)\frac{a^2 b^2}{r^2} \tag{6.32}$$

$$\sigma_\theta = \left(\frac{p_i a^2 - p_0 b^2}{b^2 - a^2}\right) + \left(\frac{p_i - p_0}{b^2 - a^2}\right)\frac{a^2 b^2}{r^2} \tag{6.33}$$

$$\sigma_z = 2v\left(\frac{p_0 a^2 - p_i b^2}{b^2 - a^2}\right) \tag{6.34}$$

It is observed that the values of σ_r and σ_θ are identical to those in the plane stress case. But in the plane stress case, $\sigma_z = 0$, whereas in the plane strain case, σ_z has a constant value given by Eq. (6.34).

6.9 Rotating Discs of Uniform Thickness

The equation of equilibrium with body force component given by (Eq. 2.46)

$$\frac{d\sigma_r}{dr} + \left(\frac{\sigma_r - \sigma_\theta}{r}\right) + F_r = 0 \tag{a}$$

is used to treat the case of a rotating disc, provided that the centrifugal "inertia force" is included as a body force. It is assumed that the stresses induced by rotation are distributed symmetrically about the axis of rotation and also independent of disc thickness. Thus, application of Eq. (a), with the body force per unit volume F_r equated to the centrifugal force $\rho w^2 r$, yields

$$\frac{d\sigma_r}{dr} + \left(\frac{\sigma_r - \sigma_\theta}{r}\right) + \rho w^2 r = 0 \tag{6.35}$$

where ρ is the mass density and w the constant angular speed of the disc in rad/sec. Equation (6.35) can be written as

$$\frac{d}{dr}(r\sigma_r) - \sigma_\theta + \rho w^2 r^2 = 0 \tag{6.36}$$

But the strain components are given by

$$\varepsilon_r = \frac{du}{dr} \text{ and } \varepsilon_\theta = \frac{u}{r} \tag{6.37}$$

From Hooke's law, with $\sigma_z = 0$

$$\varepsilon_r = \frac{1}{E}(\sigma_r - v\sigma_\theta) \tag{6.38}$$

$$\varepsilon_\theta = \frac{1}{E}(\sigma_\theta - v\sigma_r) \tag{6.39}$$

From Eq. (6.37),

$$u = r\varepsilon_\theta$$

$$\frac{du}{dr} = \varepsilon_r = \frac{d}{dr}(r\varepsilon_\theta)$$

Using Hooke's law, we can write Eq. (6.38) as

$$\frac{1}{E}(\sigma_r - v\sigma_\theta) = \frac{1}{E}\left[\frac{d}{dr}(r\sigma_\theta - vr\sigma_r)\right] \tag{6.40}$$

Let

$$r\sigma_r = y. \tag{6.41}$$

Then from Eq. (6.36)

$$\sigma_\theta = \frac{dy}{dr} + \rho w^2 r^2 \tag{6.42}$$

Substituting these in Eq. (6.40), we obtain

$$r^2\frac{d^2y}{dr^2} + r\frac{dy}{dr} - y + (3 + v) + \rho w^2 r^3 = 0$$

The solution of the above differential equation is

$$y = Cr + C_1\left(\frac{1}{r}\right) - \left(\frac{3+v}{8}\right)\rho w^2 r^3 \tag{6.43}$$

From Eqs. (6.41) and (6.42), we obtain

$$\sigma_r = C + C_1\left(\frac{1}{r^2}\right) - \left(\frac{3+v}{8}\right)\rho w^2 r^2 \tag{6.44}$$

$$\sigma_\theta = C - C_1\left(\frac{1}{r^2}\right) - \left(\frac{1+3v}{8}\right)\rho w^2 r^2 \tag{6.45}$$

The constants of integration are determined from the boundary conditions.

6.9.1 Solid Disc

For a solid disc, it is required to take $C_1 = 0$, otherwise, the stresses σ_r and σ_θ become infinite at the centre. The constant C is determined from the condition at the periphery ($r = b$) of the disc. If there are no forces applied, then

$$(\sigma_r)_{r=b} = C - \left(\frac{3+v}{8}\right)\rho w^2 b^2 = 0$$

Therefore,

$$C = \left(\frac{3+v}{8}\right)\rho w^2 b^2 \tag{6.46}$$

Hence, Eqs. (6.44) and (6.45) become,

$$\sigma_r = \left(\frac{3+v}{8}\right)\rho w^2 (b^2 - r^2) \tag{6.47}$$

$$\sigma_\theta = \left(\frac{3+v}{8}\right)\rho w^2 b^2 - \left(\frac{1+3v}{8}\right)\rho w^2 r^2 \tag{6.48}$$

The stresses attain their maximum values at the centre of the disc, i.e., at $r = 0$. Therefore,

$$\sigma_r = \sigma_\theta = \left(\frac{3+v}{8}\right)\rho w^2 b^2 \tag{6.49}$$

6.9.2 Circular Disc with a Hole

Let $a = $ Radius of the hole.

If there are no forces applied at the boundaries a and b, then.
$(\sigma_r)_{r=a} = 0$, $(\sigma_r)_{r=b} = 0$.
from which we find that

$$C = \left(\frac{3+v}{8}\right)\rho w^2 (b^2 + a^2)$$

and

$$C_1 = -\left(\frac{3+v}{8}\right)\rho w^2 a^2 b^2$$

Substituting the above in Eqs. (6.44) and (6.45), we obtain

$$\sigma_r = \left(\frac{3+v}{8}\right)\rho w^2 \left(b^2 + a^2 - \left(\frac{a^2 b^2}{r^2}\right) - r^2\right) \tag{6.50}$$

$$\sigma_\theta = \left(\frac{3+v}{8}\right)\rho w^2 \left(b^2 + a^2 + \left(\frac{a^2 b^2}{r^2}\right) - \left(\frac{1+3v}{3+v}\right)r^2\right) \tag{6.51}$$

The radial stress σ_r reaches its maximum at $r = \sqrt{ab}$, where

$$(\sigma_r)_{\max} = \left(\frac{3+\nu}{8}\right)\rho w^2 (b-a)^2 \tag{6.52}$$

The maximum circumferential stress is at the inner boundary, where

$$(\sigma_\theta)_{\max} = \left(\frac{3+\nu}{4}\right)\rho w^2 \left(b^2 + \left(\frac{1-\nu}{3+\nu}\right)a^2\right) \tag{6.53}$$

The displacement u_r for all the cases considered can be calculated as below:

$$u_r = r\varepsilon_\theta = \frac{r}{E}(\sigma_\theta - \nu\sigma_r) \tag{6.54}$$

6.10   Stress Concentration

While discussing the case of simple tension and compression, it has been assumed that the bar has a prismatical form. Then for centrally applied forces, the stress at some distance from the ends is uniformly distributed over the cross-section. Abrupt changes in cross-section give rise to great irregularities in stress distribution. These irregularities are of particular importance in the design of machine parts subjected to variable external forces and to reversal of stresses. If there exists in the structural or machine element a discontinuity that interrupts the stress path, the stress at that discontinuity may be considerably greater than the nominal stress on the section; thus there is a "stress concentration" at the discontinuity. The ratio of the maximum stress to the nominal stress on the section is known as the 'Stress Concentration Factor'. Thus, the expression for the maximum normal stress in a centrically loaded member becomes

$$\sigma = K\left(\frac{P}{A}\right) \tag{6.55}$$

where A is either gross or net area (area at the reduced section), K = stress concentration factor and P is the applied load on the member. In Fig. 6.8a, b, c, the type of discontinuity is shown and in Fig. 6.8d, e, f, the approximate distribution of normal stress on a transverse plane is shown.

Stress concentration is a matter, which is frequently overlooked by designers. The high stress concentration found at the edge of a hole is of great practical importance. As an example, holes in ships decks may be mentioned. When the hull of a ship is bent, tension or compression is produced in the decks and there is a high stress concentration at the holes. Under the cycles of stress produced by waves, fatigue of the metal at the overstressed portions may result finally in fatigue cracks.

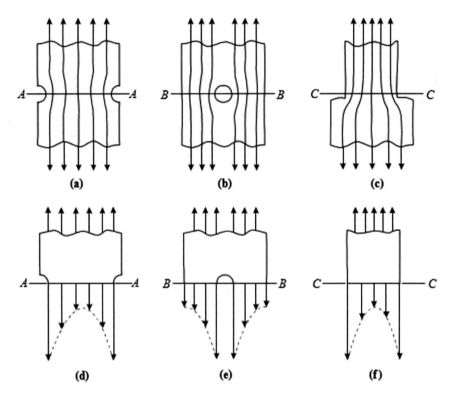

Fig. 6.8   Irregularities in stress distribution

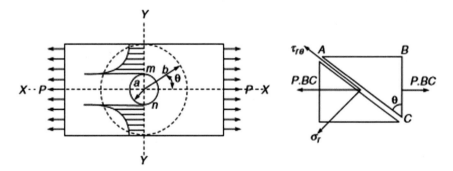

Fig. 6.9   Plate with a circular hole

6.11   The Effect of Circular Holes on Stress Distributions in Plates

Consider a plate subjected to a uniform tensile stress p as shown in Fig. 6.9. The plate thickness is small in comparison to its width and length so that we can treat this problem as a plane stress case. Let a hole of radius "a" is drilled in the middle of the plate as shown in the figure. This hole will disturb the stress field in the neighbourhood of the hole. But from St. Venant's principle, it can be assumed that any disturbance in the uniform stress field will be localized to an area within a circle of radius "b". Beyond this circle, it is expected that the stresses to be effectively the same as in the plate without the hole.

Now consider the equilibrium of an element ABC at $r = b$ and angle θ with respect to x-axis.

$$\therefore \sigma_r = p\, BC \left(\frac{\cos \theta}{AC} \right)$$

$$= p \cos^2 \theta$$

$$\therefore \sigma_r = \frac{p}{2}(1 + \cos 2\theta) \tag{6.56}$$

and

$$\tau_{r\theta} = -p.BC \left(\frac{\sin \theta}{AC} \right)$$

$$= -p\, \sin \theta \cos \theta$$

$$\therefore \tau_{r\theta} = -\frac{p}{2} \sin 2\theta \tag{6.57}$$

These stresses, acting around the outside of the ring having the inner and outer radii $r = a$ and $r = b$, give a stress distribution within the ring which may be regarded as consisting of two parts.

(a)   A constant radial stress $\frac{p}{2}$ at radius b. This condition corresponds to the ordinary thick cylinder theory, and stresses σ_r' and σ_θ' at radius r is given by

$$\sigma_r' = A + \left(\frac{B}{r^2} \right) \quad \text{and} \quad \sigma_\theta' = A - \left(\frac{B}{r^2} \right)$$

Constants A and B are given by boundary conditions,

(i)   At $r = a$, $\sigma_r = 0$
(ii) At $r = b$, $\sigma_r = \frac{p}{2}$

On substitution and evaluation, we get

$$\sigma_r' = \frac{pb^2}{2(b^2 - a^2)}\left(1 - \frac{a^2}{r^2}\right)$$

$$\sigma_\theta' = \frac{pb^2}{2(b^2 - a^2)}\left(1 + \frac{a^2}{r^2}\right)$$

(b)   The second part of the stress σ_r'' and σ_θ'' are functions of θ. The boundary conditions for this are:

$$\sigma_r'' = \frac{p}{2}\cos 2\theta \quad \text{for } r = b$$

$$\tau_{r\theta}' = -\left(\frac{p}{2}\right)\sin 2\theta \quad \text{for } r = b$$

These stress components may be derived from a stress function of the form,

$$\phi = f(r)\cos 2\theta$$

because with

$$\sigma_r'' = \left(\frac{1}{r^2}\right)\frac{\partial^2 \phi}{\partial \theta^2} + \left(\frac{1}{r}\right)\frac{\partial \phi}{\partial r}$$

$$\text{and } \sigma_\theta'' = \left(\frac{1}{r^2}\right)\frac{\partial \phi}{\partial \theta} - \left(\frac{1}{r}\right)\frac{\partial^2 \phi}{\partial r \partial \theta}$$

Now, the compatibility equation is given by,

$$\left(\frac{\partial^2}{\partial r^2} + \frac{1}{r}\frac{\partial}{\partial r} + \frac{1}{r^2}\frac{\partial^2}{\partial \theta^2}\right)f(r)\cos 2\theta = 0$$

But

$$\frac{\partial^2}{\partial r^2}f(r)\cos 2\theta + \frac{1}{r}\frac{\partial}{\partial r}f(r)\cos 2\theta + \frac{1}{r^2}\frac{\partial^2}{\partial \theta^2}f(r)\cos 2\theta$$

$$= \cos 2\theta \left\{\frac{\partial^2}{\partial r^2}f(r) + \frac{1}{r}\frac{\partial}{\partial r}f(r) - \frac{4}{r^2}f(r)\right\}$$

Therefore, the compatibility condition reduces to

$$\cos 2\theta \left\{\frac{\partial}{\partial r^2} + \frac{1}{r}\frac{\partial}{\partial r} - \frac{4}{r^2}\right\}^2 f(r) = 0$$

As $\cos 2\theta$ is not in general zero, we have

$$\left\{\frac{\partial}{\partial r^2} + \frac{1}{r}\cdot\frac{\partial}{\partial r} - \frac{4}{r^2}\right\}^2 f(r) = 0$$

We find the following ordinary differential equation to determine $f(r)$.
i.e.,

$$\left\{\frac{d^2}{dr^2} + \frac{1}{r}\frac{d}{dr} - \frac{4}{r^2}\right\}\left\{\frac{d^2 f}{dr^2} + \frac{1}{r}\frac{df}{dr} - \frac{4f}{r^2}\right\} = 0$$

On simplification, we get

$$\frac{d^4 f}{dr^4} + \frac{2}{r}\frac{d^3 f}{dr^3} - \frac{9}{r^2}\frac{d^2 f}{dr^2} + \frac{9}{r^3}\frac{df}{dr} = 0$$

This is an ordinary differential equation, which can be reduced to a linear differential equation with constant coefficients by introducing a new variable "t" such that $r = e^t$.

Also,

$$\frac{df}{dr} = \frac{df}{dt}\frac{dt}{dr} = \frac{1}{r}\frac{df}{dt}$$

$$\frac{d^2 f}{dr^2} = \frac{1}{r^2}\left(\frac{d^2 f}{dt^2} - \frac{df}{dt}\right)$$

$$\frac{d^3 f}{dr^3} = \frac{1}{r^3}\left(\frac{d^3 f}{dt^3} - 3\frac{d^2 f}{dt^2} + 2\frac{df}{dt}\right)$$

$$\frac{d^4 f}{dr^4} = \frac{1}{r^4}\left(\frac{d^4 f}{dt^4} - 6\frac{d^3 f}{dt^3} + 11\frac{d^2 f}{dt^2} - 6\frac{df}{dt}\right)$$

on substitution, we get.

$$\frac{1}{r^4}\left(\frac{d^4 f}{dt^4} - 6\frac{d^3 f}{dt^3} + 11\frac{d^2 f}{dt^2} - 6\frac{df}{dt}\right) + \frac{2}{r^4}\left(\frac{d^3 f}{dt^3} - 3\frac{d^2 f}{dt^2} + 2\frac{df}{dt}\right)$$

$$- \frac{9}{r^4}\left(\frac{d^2 f}{dt^2} - \frac{df}{dt}\right) + \frac{9}{r^4}\left(\frac{df}{dt}\right) = 0$$

$$\text{or} \quad \frac{d^4 f}{dt^4} - 4\frac{d^3 f}{dt^3} - 4\frac{d^2 f}{dt^2} + 16\frac{df}{dt} = 0.$$

Let $\frac{df}{dt} = m$

$$\therefore m^4 - 4m^3 - 4m^2 + 16m = 0$$

$$m^3(m - 4) - 4m(m - 4) = 0$$

$$\therefore (m - 4)(m^3 - 4m) = 0$$

$$\text{or } m(m^2 - 4)(m - 4) = 0$$

$$\therefore m = 0, \ m = \pm 2, \ m = 4$$

$$\therefore f(r) = A\,e^{2t} + B\,e^{4t} + C\,e^{-2t} + D$$

$$\therefore f(r) = A\,r^2 + B\,r^4 + \frac{C}{r^2} + D$$

The stress function may now be written in the form:

$$\phi = \left(A\,r^2 + B\,r^4 + \frac{C}{r^2} + D \right) \cos 2\theta$$

The stress components σ_r'' and σ_θ'' may now expressed as

$$\sigma_r'' = \left(\frac{1}{r^2} \right) \frac{\partial^2 \phi}{\partial \theta^2} + \left(\frac{1}{r} \right) \frac{\partial \phi}{\partial r}$$

$$\therefore \sigma_r'' = -\left(2A + \frac{6C}{r^4} + \frac{4D}{r^2} \right) \cos 2\theta$$

and

$$\sigma_\theta'' = \frac{\partial^2 \phi}{\partial r^2}$$

$$\therefore \sigma_\theta'' = \left(2A + 12Br^2 + \frac{6C}{r^4} \right) \cos 2\theta$$

and

$$\tau_{r\theta}'' = \left(\frac{1}{r^2} \right) \frac{\partial \phi}{\partial \theta} - \left(\frac{1}{r^2} \right) \frac{\partial^2 \phi}{\partial r \partial \theta}$$

$$\therefore \tau_{r\theta}'' = \left(2A + 6Br^2 - \frac{6C}{r^2} - \frac{2D}{r^2} \right) \sin 2\theta$$

The boundary conditions are,

(a)   At $r = a$, $\sigma_r'' = 0$
(b)   At $r = b$, $\sigma_r'' = \frac{p}{2} \cos 2\theta$
(c)   At $r = b$, $\tau_{r\theta}'' = -\frac{p}{2} \sin 2\theta$
(d)   At $r = a$, $\tau_{r\theta}'' = 0$.

Therefore, we have on substitution in stress components,

$$2A + \frac{6C}{a^4} + \frac{4D}{a^2} = 0$$

$$2A + \frac{6C}{b^4} + \frac{4D}{b^2} = -\frac{p}{2}$$

$$2A + 6Ba^2 - \frac{6C}{a^4} - \frac{2D}{a^2} = 0$$

$$2A + 6Bb^2 - \frac{6C}{b^4} - \frac{2D}{b^2} = -\frac{p}{2}$$

Solving the above, we get

$$B = -\left[\frac{pa^2b^2}{2(a^2 - b^2)^3}\right]$$

If "a" is very small in comparison to b, we may write $B \cong 0$.
Now, taking approximately,

$$D = \frac{a^2 p}{2}$$

$$C = -\left(\frac{a^4 p}{4}\right)$$

$$A = -\left(\frac{p}{4}\right)$$

Therefore, the total stress can be obtained by adding part (a) and part (b). Hence, we have

$$\sigma_r = \sigma_r' + \sigma_r'' = \frac{p}{2}\left(1 - \frac{a^2}{r^2}\right) + \frac{p}{2}\left(1 + \frac{3a^4}{r^4} - 4\frac{a^2}{r^2}\right)\cos 2\theta \qquad (6.58)$$

$$\sigma_\theta = \sigma_\theta' + \sigma_\theta'' = \frac{p}{2}\left(1 + \frac{a^2}{r^2}\right) - \frac{p}{2}\left(1 + \frac{3a^4}{r^4}\right)\cos 2\theta \qquad (6.59)$$

and

$$\tau_{r\theta} = \tau_{r\theta}'' = -\frac{p}{2}\left(1 - \frac{3a^4}{r^4} + \frac{2a^2}{r^2}\right)\sin 2\theta \qquad (6.60)$$

Now, At $r = a$, $\sigma_r = 0$

$$\therefore \sigma_r = p - 2p\cos 2\theta$$

When $\theta = \frac{\pi}{2}$ or $\frac{3\pi}{2}$

$$\sigma_\theta = 3p$$

When $\theta = 0$ or $\theta = \pi$

$$\sigma_\theta = -p$$

Fig. 6.10 Plate subjected to
stresses in two directions

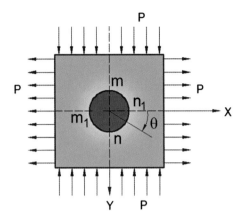

Therefore, we find that at points m and n, the stress σ_θ is three times the intensity
of applied stress. The peak stress $3p$ rapidly dies down as we move from $r = a$ to $r
= b$ since at $\theta = \frac{\pi}{2}$

$$\sigma_\theta = \frac{p}{2}\left(2 + \frac{a^2}{r^2} + \frac{3a^4}{r^4}\right)$$

which rapidly approaches p as r increases.

From the above, one can conclude that the effect of drilling a hole in highly
stressed element can lead to serious weakening.

Now, having the solution for tension or compression in one direction, the solution for tension or compression in two perpendicular directions can be obtained by
superposition. However, by taking, for instance, tensile stresses in two perpendicular
directions equal to p, we find at the boundary of the hole a tensile stress $\sigma_\theta = 2p$.
Also, by taking a tensile stress p in the x-direction and compressive stress $-p$ in the
y-direction as shown in Fig. 6.10, we obtain the case of pure shear.

Therefore, the tangential stresses at the boundary of the hole are obtained from
Eqs. (6.58), (6.59) and (6.60),
i.e.

$$\sigma_\theta = p - 2p\cos 2\theta - [p - 2p\cos(2\theta - \pi)]$$

For $\theta = \frac{\pi}{2}$ or $\theta = \frac{3\pi}{2}$ that is, at the points n and m,

$$\sigma_\theta = 4p$$

For $\theta = 0$ or $\theta = \pi$, that is, at the points n_1 and m_1, $\sigma_\theta = -4p$.

Hence for a large plate under pure shear, the maximum tangential stress at the
boundary of the hole is four times the applied pure shear stress.

6.12 Bars with Large Initial Curvature

There are practical cases of bars, such as hooks, links and rings, which have large initial curvature. In such a case, the dimensions of the cross-section are not very small in comparison with either the radius of curvature or with the length of the bar. The treatment that follows is based on the theory due to Winkler and Bach.

6.13 Winkler–Bach Theory

Assumptions in Winkler's Theory.

1. Transverse sections which are plane before bending remain plane even after bending.
2. Longitudinal fibres of the bar, parallel to the central axis exert no pressure on each other.
3. All cross-sections possess a vertical axis of symmetry lying in the plane of the centroidal axis passing through C (Fig. 6.11).
4. The beam is subjected to end couples M. The bending moment vector is normal throughout the plane of symmetry of the beam.

Winkler–Bach Formula to Determine Bending Stress or Normal Stress (Also known as Circumferential Stress).

Consider a curved beam of constant cross-section, subjected to pure bending produced by couples M applied at the ends. On the basis of plane sections remaining plane, we can state that the total deformation of a beam fibre obeys a linear law, as the beam element rotates through small angle $\Delta d\theta$. But the tangential strain ε_θ does not follow a linear relationship.

The deformation of an arbitrary fibre, $gh = \varepsilon_c R d\theta + y \Delta d\theta$.

where ε_c denotes the strain of the centroidal fibre.

But the original length of the fibre $gh = (R + y)\, d\theta$.

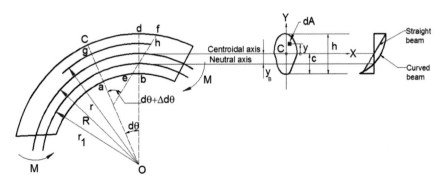

Fig. 6.11 Beam with large initial curvature

Therefore, the tangential strain in the fibre $gh = \varepsilon_\theta = \frac{[\varepsilon_c R d\theta + y \Delta d\theta]}{(R+y)d\theta}$.
Using Hooke's law, the tangential stress acting on area dA is given by

$$\sigma_\theta = \frac{\varepsilon_c R + y(\Delta d\theta/d\theta)}{(R+y)} E \tag{6.61}$$

Let angular strain $\frac{\Delta d\theta}{d\theta} = \lambda$.
Hence, Eq. (6.61) becomes

$$\sigma_\theta = \frac{\varepsilon_c R + y\lambda}{(R+y)} E \tag{6.62}$$

Adding and subtracting $\varepsilon_c y$ in the numerator of Eq. (6.62), we get,

$$\sigma_\theta = \frac{\varepsilon_c R + y\lambda + \varepsilon_c y - \varepsilon_c y}{(R+y)} E$$

Simplifying, we get

$$\sigma_\theta = \left[\varepsilon_c + (\lambda - \varepsilon_c) \frac{y}{(R+y)} \right] E \tag{6.63a}$$

The beam section must satisfy the conditions of static equilibrium,
$F_z = 0$ and $M_x = 0$, respectively:

$$\therefore \int \sigma_\theta dA = 0 \quad \text{and} \quad \int \sigma_\theta y dA = M \tag{6.63b}$$

Substituting the above boundary conditions (6.63b) in (6.63a), we get

$$0 = \int \left[\varepsilon_c + (\lambda - \varepsilon_c) \frac{y}{(R+y)} \right] dA$$

$$\text{or} \int \varepsilon_c dA = -(\lambda - \varepsilon_c) \int \frac{y}{(R+y)} dA$$

$$\text{or} \ \varepsilon_c \int dA = -(\lambda - \varepsilon_c) \int \frac{y}{(R+y)} dA \tag{6.63c}$$

Also,

$$M = \left[\varepsilon_c \int y dA + (\lambda - \varepsilon_c) \int \frac{y^2}{(R+y)} dA \right] E \tag{6.63d}$$

Here $\int dA = A$, and since y is measured from the centroidal axis, $\int y dA = 0$.
Let $\int \frac{y}{(R+y)} dA = -mA$.
Or

$$m = -\frac{1}{A} \int \frac{y}{(R+y)} \mathrm{d}A$$

Therefore,

$$\int \frac{y^2}{(R+y)} \mathrm{d}A = \int \left(y - \frac{Ry}{(R+y)} \right) \mathrm{d}A$$

$$= \int y \,\mathrm{d}A - \int \frac{Ry}{(R+y)} \mathrm{d}A$$

$$= 0 - R[-mA]$$

$$\therefore \int \frac{y^2}{(R+y)} \mathrm{d}A = mRA$$

Substituting the above values in (6.63c) and (6.63d), we get,

$$\varepsilon_c = (\lambda - \varepsilon_c)m$$

and $M = E\,(\lambda - \varepsilon_c)\,mAR$.
From the above, we get

$$\varepsilon_c = \frac{M}{AER} \text{ and } \lambda = \frac{1}{AE}\left(\frac{M}{R} + \frac{M}{mR} \right) \tag{6.63e}$$

Substitution of the values of Eqs. (6.63e) into (6.63a) gives an expression for the tangential stress in a curved beam subject to pure bending.
Therefore,

$$\sigma_\theta = \frac{M}{AR}\left[1 + \frac{y}{m(R+y)} \right] \tag{6.64}$$

The above expression for σ_θ is generally known as the "Winkler–Bach formula". The distribution of stress σ_θ is given by the hyperbolic (and not linear as in the case of straight beams) as shown in Fig. 6.11.

In the above expression, the quantity m is a pure number and is the property of each particular shape of the cross-section. Table 6.1 gives the formula for m for various shapes of the cross-section.

Sign Convention.

The following sign convention will be followed:

1. A bending moment M will be taken as positive when it is directed towards the concave side of the beam (or it decreases the radius of curvature), and negative if it increases the radius of curvature.
2. "y" is positive when measured towards the convex side of the beam, and negative when measured towards the concave side (or towards the centre of curvature).

Table 6.1 Value m for various shapes of cross-section

Cross-section	Formula for 'm'
A	$m = -1 + 2\left(\frac{R}{C}\right)^2 - 2\left(\frac{R}{C}\right)\sqrt{\left(\frac{R}{C}\right)^2 - 1}$
B	$m = -1 + \frac{2R}{C^2 - C_1^2}\left[\sqrt{R^2 - C_1^2} - \sqrt{R^2 - C^2}\right]$
C	$m = -1 + \frac{R}{Ah}\left\{[b_1 h + (R + C_1)(b - b_1)]\ln\left(\frac{R + C_1}{R - C}\right) - (b - b_1)h\right\}$ For Rectangular Section: $C = C_1$; $b = b_1$ For Triangular Section: $b_1 = 0$

(continued)

Table 6.1 (continued)

Cross-section	Formula for 'm'
D	$m = -1 + \frac{R}{A}[t.\ln(R + C_1) + (b - t).\ln(R - C_2) - b.\ln(R - C)]$
E	$m = -1 + \frac{R}{A}[b_1.\ln(R + C_1) + (t - b_1).\ln(R + C_3) - (b - t).\ln(R - C_2)b.\ln(R - C)]$

3. With the above sign convention, if σ_θ is positive, it denotes tensile stress while negative sign means compressive stress.

The distance between the centroidal axis ($y = 0$) and the neutral axis is found by setting the tangential stress to zero in Eq. (6.64)

$$\therefore 0 = \frac{M}{AR}\left[1 + \frac{y}{m(R+y)}\right]$$

or $1 = -\frac{y_n}{m(R - y_n)}$.

where y_n denotes the distance between axes as indicated in Fig. 6.11. From the above,

$$y_n = -\frac{mR}{(m+1)}$$

This expression is valid for pure bending only.

However, when the beam is acted upon by a normal load P acting through the centroid of cross-sectional area A, the tangential stress given by Eq. (6.64) is added to the stress produced by this normal load P. Therefore, for this simple case of superposition, we have

$$\sigma_\theta = \frac{P}{A} + \frac{M}{AR}\left[1 + \frac{y}{m(R+y)}\right] \tag{6.65}$$

As before, a negative sign is associated with a compressive load P.

6.14 Stresses in Closed Rings

Crane hook, split rings are the curved beams that are unstrained at one end or both ends. For such beams, the bending moment at any section can be calculated by applying the equations of statics directly. But for the beams having restrained or fixed ends such as a close ring, equations of equilibrium are not sufficient to obtain the solution, as these beams are statically indeterminate. In such beams, elastic behaviour of the beam is considered and an additional condition by considering the deformation of the member under given load is developed as in the case of statically indeterminate straight beam.

Now, consider a closed ring shown in Fig. 6.12a, which is subjected to a concentrated load P along a vertical diametrical plane.

The distribution of stress in upper half of the ring will be same as that in the lower half due to the symmetry of the ring. Also, the stress distribution in any one quadrant will be same as in another. However, for the purposes of analysis, let us consider a quadrant of the circular ring as shown in Fig. 6.12c, which may be considered to be

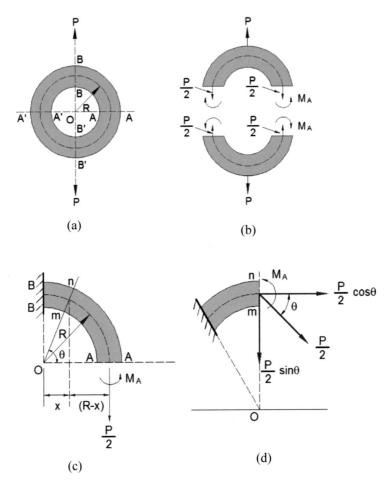

Fig. 6.12 Closed ring subjected to loads

fixed at the section BB and at section AA subjected to an axial load $\frac{P}{2}$ and bending moment M_A. Here the magnitude and the sign of the moment M_A are unknown.

Now, taking the moments of the forces that lie to the one side of the section, then we get,

$$M_{mn} = -M_A + \frac{P}{2}(R - x)$$

But from Figure, $x = R \cos \theta$

$$\therefore M_{mn} = -M_A + \frac{P}{2}(R - R \cos \theta)$$

Fig. 6.13 Section PQMN

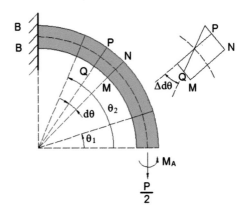

$$\therefore M_{mn} = -M_A + \frac{PR}{2}(1 - \cos\theta) \tag{a}$$

The moment M_{mn} at the section MN cannot be determined unless the magnitude of M_A is known. Resolving $\frac{P}{2}$ into normal and tangential components, we get.

Normal component, producing uniform tensile stress $= N = \frac{1}{2}P\cos\theta$.

Tangential component, producing shearing stress $= T = \frac{1}{2}P\sin\theta$.

Determination of bending moment "M_A"

Consider the elastic behaviour of the two normal sections MN and PQ, a differential distance apart. Let the initial angle $d\theta$ between the planes of these two sections change by an amount $\Delta d\theta$ when loads are applied (Fig. 6.13)..

Therefore,

$$\text{the angular strain} = \omega = \frac{\Delta d\theta}{d\theta}$$

i.e. $$\Delta d\theta = \omega\, d\theta.$$

Therefore, if we are interested in finding the total change in angle between the sections, that makes an angle θ_1 and θ_2 with the section AA, the expression $\int_{\theta_2}^{\theta_1} \omega\, d\theta$ will give that angle.

But in the case of a ring, sections AA and BB remain at right angles to each other before and after loading. Thus, the change in the angle between these planes is equal to zero. Hence

$$\int_{0}^{\pi/2} \omega\, d\theta = 0 \tag{b}$$

In straight beams, the rate of change of slope of the elastic curve is given by $\frac{d^2 y}{dx^2} = \frac{M}{EI}$. Whereas in initially curved beam, the rate of change of slope of the elastic curve is $\frac{\Delta d\theta}{R d\theta}$, which is the angle change per unit of arc length.

Now, $\frac{\Delta d\theta}{R d\theta} = \frac{\omega}{R} = \frac{M}{EI} = \frac{M_{mn}}{EI}$ for curved beams.

Or

$$\omega = \frac{R M_{mn}}{EI}$$

Substituting the above in Eq. (b), we get

$$\int\limits_{0}^{\frac{\pi}{2}} \frac{R.M_{mn}}{EI} d\theta = 0$$

since R, E and I are constants,

$$\therefore \int\limits_{0}^{\frac{\pi}{2}} M_{mn} d\theta = 0$$

From Eq. (a), substituting the value of M_{mn}, we obtain

$$-\int\limits_{0}^{\frac{\pi}{2}} M_A \, d\theta + \frac{1}{2} P R \int\limits_{0}^{\frac{\pi}{2}} d\theta - \frac{1}{2} P R \int\limits_{0}^{\frac{\pi}{2}} \cos\theta \, d\theta = 0$$

Integrating, we get

$$-M_A[\theta]_0^{\frac{\pi}{2}} + \frac{1}{2} P R[\theta]_0^{\frac{\pi}{2}} - \frac{1}{2} P R[\sin\theta]_0^{\frac{\pi}{2}} = 0$$

$$-M_A\left(\frac{\pi}{2}\right) + \frac{1}{2} P R\left(\frac{\pi}{2}\right) - \frac{1}{2} P R\left(\sin\frac{\pi}{2}\right) = 0$$

Thus $M_A = \frac{PR}{2}\left(1 - \frac{2}{\pi}\right)$.

Therefore, knowing M_A, the moment at any section such as MN can be computed and then the normal stress can be calculated by curved beam formula at any desired section.

6.15 Numerical Examples

Example 6.1 Given the following stress function.

$$\phi = \frac{P}{\pi} r\theta \cos\theta$$

Determine the stress components σ_r, σ_θ and $\tau_{r\theta}$.

Solution: The stress components, by definition of φ, are given as follows

$$\sigma_r = \left(\frac{1}{r}\right)\frac{\partial\phi}{\partial r} + \left(\frac{1}{r^2}\right)\frac{\partial^2\phi}{\partial\theta^2} \qquad\qquad (i)$$

$$\sigma_\theta = \frac{\partial^2\phi}{\partial r^2} \qquad\qquad (ii)$$

$$\tau_{r\theta} = \left(\frac{1}{r^2}\right)\frac{\partial\phi}{\partial\theta} - \left(\frac{1}{r}\right)\frac{\partial^2\phi}{\partial r\partial\theta} \qquad\qquad (iii)$$

The various derivatives are as follows:

$$\frac{\partial\phi}{\partial r} = \frac{P}{\pi}\theta\cos\theta$$

$$\frac{\partial^2\phi}{\partial r^2} = 0$$

$$\frac{\partial\phi}{\partial\theta} = \frac{P}{\pi}r(-\theta\sin\theta + \cos\theta)$$

$$\frac{\partial^2\phi}{\partial\theta^2} = -\frac{P}{\pi}r(\theta\cos\theta + 2\sin\theta)$$

$$\frac{\partial^2\phi}{\partial r\partial\theta} = \frac{P}{\pi}(-\theta\sin\theta + \cos\theta)$$

Substituting the above values in Eqs. (i), (ii) and (iii), we get

$$\sigma_r = \left(\frac{1}{r}\right)\frac{P}{\pi}\theta\cos\theta - \left(\frac{1}{r^2}\right)\frac{P}{\pi}r(\theta\cos\theta + 2\sin\theta)$$

$$= \left(\frac{1}{r}\right)\frac{P}{\pi}\theta\cos\theta - \left(\frac{1}{r}\right)\frac{P}{\pi}\theta\cos\theta - \left(\frac{1}{r}\right)\frac{P}{\pi}2\sin\theta$$

$$\therefore \sigma_r = -\frac{2}{r}\frac{P}{\pi}\sin\theta$$

$$\sigma_\theta = \frac{\partial^2\varphi}{\partial r^2} = 0$$

$$\tau_{r\theta} = \left(\frac{1}{r^2}\right)\frac{P}{\pi}r(-\theta\sin\theta + \cos\theta) - \left(\frac{1}{r}\right)\frac{P}{\pi}(-\theta\sin\theta + \cos\theta)$$

$$\therefore \tau_{r\theta} = 0$$

Therefore, the stress components are

$$\sigma_r = -\left(\frac{2}{r}\right)\frac{P}{\pi}\sin\theta$$

$$\sigma_\theta = 0$$

$$\tau_{r\theta} = 0$$

Example 6.2 A thick cylinder of inner radius 10 cm and outer radius 15 cm is subjected to an internal pressure of 12 MPa. Determine the radial and hoop stresses in the cylinder at the inner and outer surfaces.

   Solution: The radial stress in the cylinder is given by

$$\sigma_r = \left(\frac{p_i a^2 - p_o b^2}{b^2 - a^2}\right) - \left(\frac{p_i - p_o}{b^2 - a^2}\right)\frac{a^2 b^2}{r^2}$$

The hoop stress in the cylinder is given by

$$\sigma_\theta = \left(\frac{p_i a^2 - p_o b^2}{b^2 - a^2}\right) + \left(\frac{p_i - p_o}{b^2 - a^2}\right)\frac{a^2 b^2}{r^2}$$

   As the cylinder is subjected to internal pressure only, the above expressions reduce to

$$\sigma_r = \left(\frac{p_i a^2}{b^2 - a^2}\right) - \left(\frac{p_i}{b^2 - a^2}\right)\frac{a^2 b^2}{r^2}$$

and

$$\sigma_\theta = \left(\frac{p_i a^2}{b^2 - a^2}\right) + \left(\frac{p_i}{b^2 - a^2}\right)\frac{a^2 b^2}{r^2}$$

Stresses at inner face of the cylinder (i.e. at r = 10 cm):

$$\text{Radial stress} = \sigma_r = \left[\frac{12 \times (0.1)^2}{(0.15)^2 - (0.1)^2}\right] - \left[\frac{(0.15)^2(0.1)^2}{(0.1)^2}\right]\left[\frac{12}{(0.15)^2 - (0.1)^2}\right]$$

$$= 9.6 - 21.6$$

   or $\sigma_r = -12$ MPa.

$$\text{Hoop stress} = \sigma_\theta = \left[\frac{12 \times (0.1)^2}{(0.15)^2 - (0.1)^2}\right] + \left[\frac{12}{(0.15)^2 - (0.1)^2}\right]\left[\frac{(0.15)^2(0.1)^2}{(0.1)^2}\right]$$

$$= 9.6 + 21.6$$

   or $\sigma_\theta = 31.2$ MPa.

Stresses at outer face of the cylinder (i.e. at r = 15 cm):

$$\text{Radial stress} = \sigma_r = \left[\frac{12 \times (0.1)^2}{(0.15)^2 - (0.1)^2}\right] - \left[\frac{12}{(0.15)^2 - (0.1)^2}\right]\left[\frac{(0.1)^2(0.15)^2}{(0.15)^2}\right]$$

$$\sigma_r = 0$$

$$\text{Hoop stress} = \sigma_\theta = \left[\frac{12 \times (0.1)^2}{(0.15)^2 - (0.1)^2}\right] + \left[\frac{(0.1)^2(0.15)^2}{(0.15)^2}\right]\left[\frac{12}{(0.15)^2 - (0.1)^2}\right]$$

$$= 9.6 + 9.6$$

$$\text{or } \sigma_\theta = 19.2\,\text{MPa}.$$

Example 6.3 A steel tube, which has an outside diameter of 10 cm and inside diameter of 5 cm, is subjected to an internal pressure of 14 MPa and an external pressure of 5.5 MPa. Calculate the maximum hoop stress in the tube.

Solution: The maximum hoop stress occurs at $r = a$.
Therefore,

$$\text{maximum hoop stress} = (\sigma_\theta)_{\text{max}} = \left[\frac{p_i a^2 - p_0 b^2}{b^2 - a^2}\right] + \left[\frac{p_i - p_0}{b^2 - a^2}\right]\left[\frac{a^2 b^2}{a^2}\right]$$

$$= \left[\frac{p_i a^2 - p_0 b^2}{b^2 - a^2}\right] + \left[\frac{p_i - p_0}{b^2 - a^2}\right]b^2$$

$$= \frac{p_i a^2 - p_0 b^2 + p_i b^2 - p_0 b^2}{b^2 - a^2}$$

$$(\sigma_\theta)_{\text{max}} = \frac{p_i(a^2 + b^2) - 2p_0 b^2}{b^2 - a^2}$$

Therefore,

$$(\sigma_\theta)_{\text{max}} = \frac{14\left[(0.05)^2 + (0.1)^2\right] - 2 \times 5.5 \times (0.1)^2}{(0.1)^2 - (0.05)^2}$$

Or $(\sigma_\theta)_{\text{max}} = 8.67\,\text{MPa}.$

Example 6.4 A steel cylinder which has an inside diameter of 1 m is subjected to an internal pressure of 8 MPa. Calculate the wall thickness if the maximum shearing stress is not to exceed 35 MPa.

Solution: The critical point lies on the inner surface of the cylinder, i.e. at $r = a$.
We have,

$$\text{radial stress} = \sigma_r = \left[\frac{p_i a^2 - p_0 b^2}{b^2 - a^2}\right] - \left[\frac{p_i - p_0}{b^2 - a^2}\right]\frac{a^2 b^2}{r^2}$$

At $r = a$ and $p_0 = 0$,

$$\sigma_r = \left[\frac{p_i a^2 - 0}{b^2 - a^2}\right] - \left[\frac{p_i - 0}{b^2 - a^2}\right]\frac{a^2 b^2}{a^2}$$

$$= \frac{p_i a^2 - p_i b^2}{b^2 - a^2}$$

$$= \frac{-p_i(b^2 - a^2)}{(b^2 - a^2)}$$

Therefore, $\sigma_r = -p_i$.

Similarly,

$$\text{Hoop stress} = \sigma_\theta = \left[\frac{p_i a^2 - p_0 b^2}{b^2 - a^2}\right] + \left[\frac{p_i - p_0}{b^2 - a^2}\right]\frac{a^2 b^2}{r^2}$$

At $r = a$ and $p_0 = 0$,

$$\sigma_\theta = \left[\frac{p_i a^2 - 0}{b^2 - a^2}\right] + \left[\frac{p_i - 0}{b^2 - a^2}\right]\frac{a^2 b^2}{a^2}$$

$$\sigma_\theta = \frac{p_i(a^2 + b^2)}{(b^2 - a^2)}$$

Here the maximum and minimum stresses are.

$$\sigma_3 = -p_i \text{ and } \sigma_1 = \sigma_\theta.$$

But the maximum shear stress $= \tau_{max} = \frac{1}{2}(\sigma_1 - \sigma_3)$.
i.e.

$$\tau_{max} = \frac{1}{2}\left[\frac{p_i(a^2 + b^2)}{(b^2 - a^2)} + p_i\right]$$

$$= \frac{1}{2}\left[\frac{p_i a^2 + p_i b^2 + p_i b^2 - p_i a^2}{(b^2 - a^2)}\right]$$

$$35 = \frac{p_i b^2}{(b^2 - a^2)}$$

i.e.

$$35 = \frac{8 \times b^2}{(b^2 - a^2)}$$

$$35b^2 - 35a^2 = 8b^2$$
$$35b^2 - 8b^2 = 35a^2$$
$$35b^2 - 8b^2 = 35(0.5)^2$$

Fig. 6.14 Loaded circular link

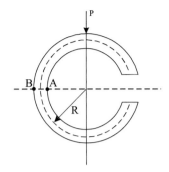

Therefore, $b = 0.5693$.
If t is the thickness of the cylinder, then.
$b = 0.5 + t = 0.5693$.
$\therefore t = 0.0693$ m or 69.3 mm.

Example 6.5 The circular link shown in Fig. 6.14 has a circular cross-section 3 cm in diameter. The inside diameter of the ring is 4 cm. The load P is 1000 kg. Calculate the stress at A and B. Compare the values with those found by the straight beam formula. Assume that the material is not stressed above its elastic strength.

Solution:

$$\text{Cross - sectional area} = A = \frac{\pi}{4}(3)^2 = 7.06 \text{ cm}^2.$$

For circular cross-section m is given by

$$m = -1 + 2\left(\frac{R}{c}\right)^2 - 2\left(\frac{R}{c}\right)\sqrt{\left(\frac{R}{c}\right)^2 - 1}$$

Here $R = 2 + 1.5 = 3.5$ cm.

$c = 1.5$ cm. (Refer Table 6.1).
Therefore,

$$m = -1 + 2\left(\frac{3.5}{1.5}\right)^2 - 2\left(\frac{3.5}{1.5}\right)\sqrt{\left(\frac{3.5}{1.5}\right)^2 - 1}$$

$$m = 0.050$$

At section AB, the load is resolved into a load P and a bending couple whose moment is positive. The stress at A and B is considered to be the sum of the stress due to axial load P, and the stress due to the bending moment M.

Therefore, stress at point A is

$$\sigma_{\theta A} = \sigma_A = \frac{P}{A} + \frac{M}{AR}\left[1 + \frac{y_A}{m(R + y_A)}\right]$$

$$= -\frac{1000}{7.06} + \frac{(3.5 \times 1000)}{7.06 \times 3.5}\left[1 + \frac{(-1.5)}{0.050(3.5 - 1.5)}\right]$$

or $\sigma_A = -2124.65$ kg/cm^2 (compressive).
The stress at point B is given by

$$\sigma_{\theta_B} = \sigma_B = +\frac{P}{A} + \frac{M}{AR}\left[1 + \frac{y_B}{m(R + y_B)}\right]$$

$$= \frac{-1000}{7.06} + \frac{3500}{7.06 \times 3.5}\left[1 + \frac{1.5}{0.050(3.5 + 1.5)}\right]$$

$\therefore \sigma_B = 849.85$ kg/cm^2 (Tensile).

Comparison by Straight Beam Formula.
The moment of inertia of the ring cross-section about the centroidal axis is

$$I = \frac{\pi d^4}{64} = \frac{\pi(3)^4}{64} = 3.976 \text{ cm}^4$$

If the link is considered to be a straight beam, the corresponding values are

$$\sigma_A = \frac{P}{A} + \frac{My}{I}$$

$$= -\frac{1000}{7.06} + \frac{(+3500)(-1.5)}{3.976}$$

$\therefore \sigma_A = -1462.06$ kg/cm^2 (compressive).
and $\sigma_B = \frac{-1000}{7.06} + \frac{3500 \times 1.5}{3.976}$.
$\sigma_B = 1178.8$ kg/cm^2 (tensile) (Fig. 6.15).

Example 6.6 An open ring having T-Section as shown in the Fig. 6.16 is subjected to a compressive load of 10,000 kg. Compute the stresses at A and B by curved beam formula.

Solution:

$$\text{Area of the section} = A = 2 \times 10 + 2 \times 14 = 48 \text{ cm}^2$$

The value of m can be calculated from Table 6.1 by substituting $b_1 = 0$ for the unsymmetric I-section.
From Figure,

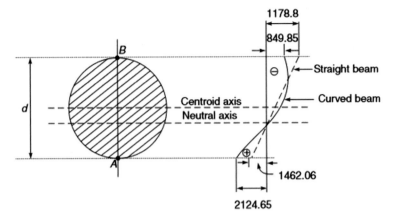

Fig. 6.15 Stresses along the cross-section

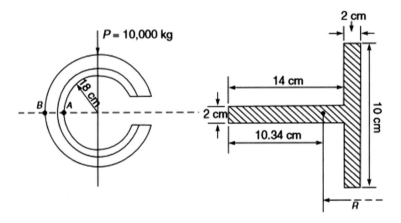

Fig. 6.16 Loaded open ring

$$R = 18 + 5.66 = 23.66\,\text{cm}.$$
$$c_1 = c_3 = 10.34\,\text{cm}.$$
$$c_2 = 3.66\,\text{cm},\ c = 5.66\,\text{cm}.$$
$$t = 2\,\text{cm}.$$
$$b_1 = 0,\ b = 10\,\text{cm}.$$

m is given by.

$$m = -1 + \frac{R}{A}\left[\begin{array}{l} b_1.\ln(R + c_1) + (t - b_1).\ln(R + c_3) \\ + (b - t).\ln(R - c_2) - b.\ln(R - c) \end{array}\right]$$

$$= -1 + \frac{23.66}{48} \left[\begin{array}{c} 0 + (2 - 0) \ln(23.66 + 10.34) \\ + (10 - 2) \ln(23.66 - 3.66) - 10 \ln(23.66 - 5.66) \end{array} \right]$$

Therefore, $m = 0.042$.
Now, stress at A,

$$\sigma_A = \frac{P}{A} + \frac{M}{AR} \left[1 + \frac{y_A}{m(R + y_A)} \right]$$

$$= -\frac{10,000}{48} + \frac{(10,000 \times 23.66)}{48 \times 23.66} \left[1 + \frac{(-5.66)}{0.042(23.66 - 5.66)} \right]$$

$\therefore \sigma_A = -1559.74 \text{ kg/cm}^2$ (compressive).
Similarly, stress at B is given by

$$\sigma_B = \frac{P}{A} + \frac{M}{AR} \left[1 + \frac{y_B}{m(R + y_B)} \right]$$

$$= -\frac{10000}{48} + \frac{10000 \times 23.66}{48 \times 23.66} \left[1 + \frac{10.34}{0.042(23.66 + 10.34)} \right]$$

$\therefore \sigma_B = 1508.52 \text{ kg/cm}^2$ (tensile).

Example 6.7 A ring shown in the Fig. 6.17, with a rectangular section is 4 cm wide and 2 cm thick. It is subjected to a load of 2000 kg. Compute the stresses at A and B and at C and D by curved beam formula.
  Solution: Area of the section $A = 4 \times 2 = 8 \text{ cm}^2$.
  The radius of curvature of the centroidal axis $= R = 4 + 2 = 6 \text{ cm}$.

From Table 6.1, the m value for trapezoidal section is given by,

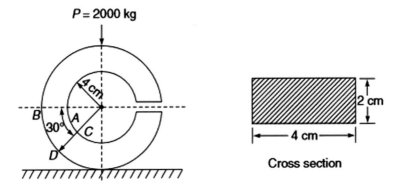

Fig. 6.17 Loaded ring with rectangular cross-section

$$m = -1 + \frac{R}{Ah}\left\{[b_1 h + (R + c_1)(b - b_1)] \ln\left(\frac{R + c_1}{R - c}\right) - (b - b_1)h\right\}$$

But for rectangular section, $c = c_1$, $b = b_1$,
Therefore

$$m = -1 + \frac{R}{Ah}\left\{[bh + (R + c)(0)] \ln\left(\frac{R + c}{R - c}\right) - (0)\right\}$$

$$m = -1 + \frac{6}{8 \times 4}\left\{[2 \times 4 + (6 + 2)(0)] \ln\left(\frac{6 + 2}{6 - 2}\right)\right\}$$

Therefore $m = 0.0397$.
Now, stress at

$$A = \sigma_A = \frac{P}{A} + \frac{M}{AR}\left[1 + \frac{y_A}{m(R + y_A)}\right]$$

$$= -\frac{2000}{8} + \frac{2000 \times 6}{8 \times 6}\left[1 + \frac{(-2)}{0.0397(6 - 2)}\right]$$

$\therefore \sigma_A = -3148.6 \, \text{kg/cm}^2$ (Compression)
Stress at

$$B = \sigma_B = \frac{P}{A} + \frac{M}{AR}\left[1 + \frac{y_B}{m(R + y_B)}\right]$$

$$= \frac{-2000}{8} + \frac{2000 \times 6}{8 \times 6}\left[1 + \frac{2}{0.0397(6 + 2)}\right]$$

Therefore, $\sigma_B = +1574.31 \, \text{kg/cm}^2$ (Tension).
To compute the stresses at C and D.

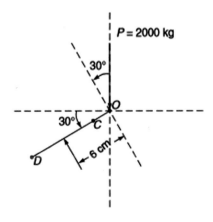

At section CD, the bending moment,

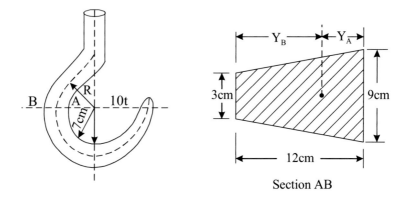

Fig. 6.18 Loaded crane hook

$$M = PR \cos 30°$$

i.e.,

$$M = 2000 \times 6 \times \cos 30°$$
$$= 10,392 \, \text{kg-cm}$$

Component of P normal to CD is given by,

$$N = P \cos 30° = 2000 \cos 30° = 1732 \, \text{kg.}$$

Therefore, stress at

$$C = \sigma_c = \frac{N}{A} + \frac{M}{AR}\left[1 + \frac{y_A}{m(R + y_A)}\right]$$
$$= \frac{-1732}{8} + \frac{10,392}{8 \times 6}\left[1 + \frac{(-2)}{0.0397(6 - 2)}\right]$$
$$\therefore \sigma_c = -2726.7 \, \text{kg/cm}^2 \, \text{(Compression)}$$

Stress at

$$D = \sigma_D = \frac{N}{A} + \frac{M}{AR}\left[1 + \frac{y_B}{m(R + y_B)}\right]$$
$$= \frac{-1732}{8} + \frac{10,392}{8 \times 6}\left[1 + \frac{2}{0.0397(6 + 2)}\right]$$

Therefore, $\sigma_D = 1363.4 \, \text{kg/cm}^2$ (Tension).

Example 6.8 The dimensions of a 10 tonne crane hook are shown in Fig. 6.18. Find the circumferential stresses σ_A and σ_B on the inside and outside fibres respectively at the section AB.

Solution: Area of the section $= A = \frac{9+3}{2} \times 12 = 72\,\text{cm}^2$.

Now, $y_A = \frac{12}{3}\left[\frac{9+2\times3}{9+3}\right] = 5\,\text{cm}$.

Therefore $y_B = (12-5) = 7\,\text{cm}$.

Radius of curvature of the centroidal axis $= R = 7 + 5 = 12\,\text{cm}$.

For trapezoidal cross-section, m is given by Table 6.1 as,

$$m = -1 + \frac{12}{72 \times 12}\left\{[(3 \times 12) + (12+7)(9-3)].\ \ln\left(\frac{12+7}{12-5}\right) - (9-3)12\right\}$$

$$\therefore m = 0.080$$

Moment $= M = PR = 10{,}000 \times 12 = 120{,}000\,\text{kg-cm}$.

Now,

Stress at

$$A = \sigma_A = \frac{P}{A} + \frac{M}{AR}\left[1 + \frac{y_A}{m(R+y_A)}\right]$$

$$= \frac{-10{,}000}{72} + \frac{120{,}000}{72 \times 12}\left[1 + \frac{(-5)}{0.08(12-5)}\right]$$

$$\therefore \sigma_A = -1240\,\text{kg/cm}^2\ (\text{Compression})$$

Stress at

$$B = \sigma_B = \frac{P}{A} + \frac{M}{AR}\left[1 + \frac{y_B}{m(R+y_B)}\right]$$

$$= \frac{-10{,}000}{72} + \frac{120{,}000}{72 \times 12}\left[1 + \frac{7}{0.08(12+7)}\right]$$

$$\therefore \sigma_B = 639.62\,\text{kg/cm}^2\ (\text{Tension})$$

Example 6.9 A circular open steel ring is subjected to a compressive force $W = 80$ kN as shown in Fig. 6.19. The cross-section of the ring is made up of an unsymmetrical I-section with an inner radius of 150 mm. Estimate the circumferential stresses developed at points A and B.

Solution:

From Table 6.1, the value of m for the above section is given by

$$m = -1$$
$$+ \frac{R}{A}[b_1 \ln(R+c_1) + (t-b_1)\ln(R+c_3) + (b-t)\ln(R-c_2) - b\ln(R-c)]$$

Hence $R = $ Radius of curvature of the centroidal axis.

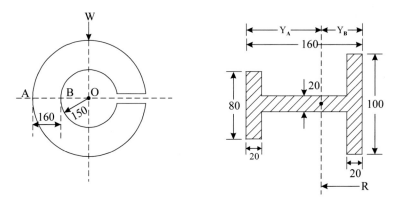

Fig. 6.19 Loaded circular ring with unsymmetrical I-section

Now, $A = 20 \times 100 + 120 \times 20 + 80 \times 20 = 6000\,\text{mm}^2$

$$y_B = \frac{(100 \times 20 \times 10) + (120 \times 20 \times 80) + (80 \times 20 \times 150)}{6000} = 75.33\,\text{mm}.$$

$$\therefore y_A = (160 - 75.33) = 84.67\,\text{mm}.$$

Also, $R = (150 + 75.33) = 225.33\,\text{mm}.$

$$\therefore m = -1 + \frac{225.33}{6000}\begin{bmatrix} 80\ln(225.33 + 84.67) + (20 - 80)\ln(225.33 + 64.67) + \\ (100 - 20)\ln(225.33 - 55.33) - 100\ln(225.33 - 75.33) \end{bmatrix}$$

$$\therefore m = 0.072.$$

Moment $= M = PR = 80 \times 1000 \times 225.33 = 1.803 \times 10^7\,\text{N} - \text{mm}.$

Now, Stress at point $B = \sigma_B = \frac{P}{A} + \frac{M}{AR}\left[1 + \frac{y_B}{m(R + y_B)}\right]$

$$\therefore \sigma_B = -\frac{80{,}000}{6000} + \frac{1.803 \times 10^7}{6000 \times 225.33}\left[1 + \frac{(-75.33)}{0.072(225.33 - 75.33)}\right]$$

$$\therefore \sigma_B = -93.02\,\text{N/mm}^2\,(\text{Compression})$$

Stress at point

$$A = \sigma_A = \frac{P}{A} + \frac{M}{AR}\left[1 + \frac{y_A}{m(R + y_A)}\right]$$

$$= -\frac{80{,}000}{6000} + \frac{1.803 \times 10^7}{6000 \times 225.33}\left[1 + \frac{84.67}{0.072(225.33 + 84.67)}\right]$$

$$\therefore \sigma_A = 50.6\,\text{N/mm}^2\,(\text{Tension})$$

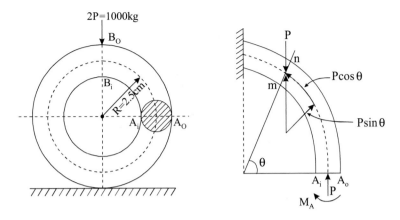

Fig. 6.20   Loaded closed ring

Hence, the resultant stresses at A and B are,
$$\sigma_A = 50.6 \ \text{N/mm}^2(\text{Tension}), \ \sigma_B = -93.02 \ \text{N/mm}^2(\text{Compression})$$

Example 6.10 Calculate the circumferential stress on inside and outside fibre of the ring at A and B, shown in Fig. 6.20. The mean diameter of the ring is 5 cm, and cross-section is circular with 2 cm diameter. Loading is within elastic limit.

Solution: For circular section, from Table 6.1

$$m = -1 + 2\left(\frac{R}{c}\right)^2 - 2\left(\frac{R}{c}\right)\sqrt{\left(\frac{R}{c}\right)^2 - 1}$$

$$= -1 + 2\left(\frac{2.5}{1}\right)^2 - 2\left(\frac{2.5}{1}\right)\sqrt{\left(\frac{2.5}{1}\right)^2 - 1}$$

$$\therefore m = 0.0435$$

We have,

$$M_A = PR\left(1 - \frac{2}{\pi}\right)$$

$$= 0.364 \ PR = 0.364 \times 2.5 \ P$$

$$\therefore M_A = 0.91 \ P$$

Now,

$$\sigma_{A_i} = -\left(\frac{P}{A}\right) + \frac{M_A}{AR}\left[1 + \frac{y_i}{m(R + y_A)}\right]$$

$$= -\left(\frac{P}{A}\right) + \frac{0.91 \ P}{A \times 2.5}\left[1 + \frac{(-1)}{0.0435(2.5 - 1)}\right]$$

$$= -\left(\frac{P}{A}\right) - 5.21\left(\frac{P}{A}\right)$$

$$\therefore \sigma_{A_i} = -6.21\left(\frac{P}{A}\right) \text{ (Compressive)}$$

$$\sigma_{A_0} = -\left(\frac{P}{A}\right) + \frac{M}{AR}\left[1 + \frac{y_o}{m(R + y_B)}\right]$$

$$= -\left(\frac{P}{A}\right) + \frac{0.91\,P}{A \times 2.5}\left[1 + \frac{1}{0.0435(2.5 + 1)}\right]$$

$$\therefore \sigma_{A_o} = 1.755\left(\frac{P}{A}\right) \text{(Tension)}$$

Similarly,

$$M_B = (M_A - PR)$$
$$= (0.364\,PR - PR) = -0.636\,PR$$
$$= -0.636 \times 2.5\,P$$
$$\therefore M_B = -1.59P$$

Now,

$$\sigma_{Bi} = \frac{M_B}{AR}\left[1 + \frac{y_i}{m(R + y_i)}\right]$$

$$= -\frac{1.59\,P}{A \times 2.5}\left[1 + \frac{(-1)}{0.0435(2.5 - 1)}\right]$$

$$\therefore \sigma_{Bi} = 9.11\left(\frac{P}{A}\right) \text{(Tension)}$$

and

$$\sigma_{Bo} = -\left(\frac{1.59\,P}{A \times 2.5}\right)\left[1 + \frac{(+1)}{0.0435(2.5 + 1)}\right]$$

$$= -4.81\left(\frac{P}{A}\right) \text{ (Compression)}$$

Now, substituting the values of $P = 500$ kg,
$A = \pi(1)^2 = 3.14159$ cm^2, above stresses can be calculated as below.

$$\sigma_{A_i} = -6.21 \times \frac{500}{\pi} = -988 \text{ kg/cm}^2$$

$$\sigma_{A_0} = 1.755 \times \frac{500}{\pi} = 279.32 \text{ kg/cm}^2$$

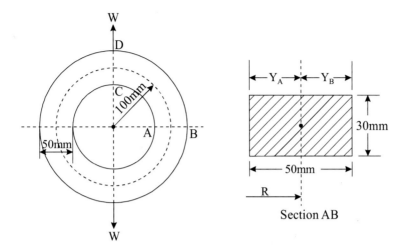

Fig. 6.21 Closed ring with rectangular cross-section

$$\sigma_{B_i} = 9.11 \times \frac{500}{\pi} = 1450 \, \text{kg/cm}^2$$

$$\sigma_{B_0} = -4.81 \times \frac{500}{\pi} = -765.54 \, \text{kg/cm}^2$$

Example 6.11 A ring of 200 mm mean diameter has a rectangular cross-section with 50 mm in the radial direction and 30 mm perpendicular to the radial direction as shown in Fig. 6.21. If the maximum tensile stress is limited to 120 N/mm², determine the tensile load that can be applied on the ring.

Solution: $R = 100$ mm, Area of cross-section $= A = 30 \times 50 = 1500$ mm².
From Table 6.1, the value of m for the rectangular section is given by

$$m = -1 + \frac{100}{1500 \times 50} \left\{ [30 \times 50 + 0] \ln\left(\frac{100+25}{100-25}\right) - 0 \right\}$$
$$\therefore m = 0.0217$$

To find M_{AB}.

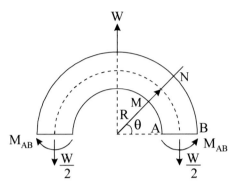

The bending moment at any section MN can be determined by

$$M_{MN} = -M_{AB} + \frac{WR}{2}(1 - \cos\theta)$$

$$\therefore At\ \theta = 0, \qquad M_{mn} = -M_{AB}$$

But

$$M_{AB} = \frac{WR}{2}\left(1 - \frac{2}{\pi}\right)$$

$$\therefore M_{AB} = \frac{W \times 100}{2}\left(1 - \frac{2}{\pi}\right) = 18.17\ W$$

Now,

$$\sigma_A = \frac{P}{A} + \frac{M_A}{AR}\left[1 + \frac{y_A}{m(R + y_A)}\right]$$

$$= \frac{W}{2A} + \frac{M_A}{AR}\left[1 + \frac{y_A}{m(R + y_A)}\right]$$

$$= \frac{W}{2 \times 1500} + \frac{(-18.17\ W)}{1500 \times 100}\left(1 + \frac{(-25)}{0.0217(100 - 25)}\right)$$

$$\therefore \sigma_A = 0.002073\ W\,(Tensile)$$

and

$$\sigma_B = \frac{P}{A} + \frac{M_A}{AR}\left[1 + \frac{y_B}{m(R + y_B)}\right]$$

$$= \frac{w}{2 \times 1500} - \frac{18.17W}{1500 \times 100}\left[1 + \frac{25}{0.0217(100 + 25)}\right]$$

$$\therefore \sigma_B = -0.00090423W\,(Compression)$$

To find stresses at C and D.
We have,

$$M_{mn} = -M_{AB} + \frac{WR}{2}(1 - \cos\theta)$$

$$\therefore At\ \theta = 90°, \quad M_{mn} = M_{CD} = -M_{AB} + \frac{WR}{2}$$

$$\therefore M_{CD} = -18.17W + W \times \frac{100}{2} = 31.83W$$

Now, stress at

$$C = \sigma_C = \frac{P}{A} + \frac{M_{CD}}{AR}\left[1 + \frac{y_C}{m(R + y_C)}\right]$$

$$= 0 + \frac{31.83W}{1500 \times 100}\left[1 + \frac{(-25)}{0.0217(100 - 25)}\right]$$

$$= -0.00305W\,(\text{Compression})$$

and stress at

$$D = \sigma_D = \frac{P}{A} + \frac{M_{CD}}{AR}\left[1 + \frac{y_D}{m(R + y_D)}\right]$$

$$= 0 + \frac{31.83W}{1500 \times 100}\left[1 + \frac{25}{0.0217(100 + 25)}\right]$$

$$\therefore \sigma_D = 0.00217W\,(\text{Tensile})$$

By comparison, the tensile stress is maximum at Point D.

$$\therefore 0.00217W = 120 \quad \therefore W = 55,299.54\,\text{N or}\,55.3\,\text{kN}$$

Example 6.12 A ring of mean diameter 100 mm is made of mild steel with 25 mm diameter. The ring is subjected to four pulls in two directions at right angles to each other passing through the centre of the ring. Determine the maximum value of the pulls if the tensile stress should not exceed 80 N/mm² (Fig. 6.22).
Solution: Here $R = 50$ mm.
From Table 6.1, the value of m for circular section is given by,

$$m = -1 + 2\left(\frac{R}{C}\right)^2 - 2\left(\frac{R}{C}\right)\sqrt{\left(\frac{R}{C}\right)^2 - 1}$$

$$= -1 + 2\left(\frac{50}{12.5}\right)^2 - 2\left(\frac{50}{12.5}\right)\sqrt{\left(\frac{50}{12.5}\right)^2 - 1}$$

$$\therefore m = 0.016$$

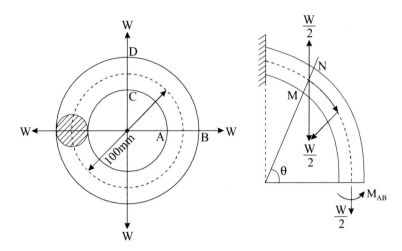

Fig. 6.22 Closed ring with circular cross-section

Area of cross-section $= A = \pi(12.5)^2 = 490.87\,\text{mm}^2$.
We have,

$$M_A = \frac{WR}{2}\left(1 - \frac{2}{\pi}\right)$$

$$= \frac{W}{2} \times 50\left(1 - \frac{2}{\pi}\right)$$

$$M_A = 9.085W$$

Now,

$$\sigma_A = \frac{P}{A} + \frac{M_A}{AR}\left[1 + \frac{y_A}{m(R + y_A)}\right]$$

$$= \frac{W}{2 \times 490.87} - \frac{9.085W}{490.87 \times 50}\left[1 + \frac{(-12.5)}{0.016(50 - 12.5)}\right]$$

$$= 0.0084W \ \text{(Tensile)}$$

$$\therefore \sigma_B = \frac{P}{A} + \left(\frac{M_A}{-AR}\right)\left[1 + \frac{y_B}{m(R + y_B)}\right]$$

$$= \frac{W}{2 \times 490.87} - \frac{9.085W}{490.87 \times 50}\left[1 + \frac{12.5}{0.016(50 + 12.5)}\right]$$

$$\therefore \sigma_B = -0.00398W \ \text{(Compression)}$$

Also,

$$M_{CD} = (M_A - PR) = \left(-9.085 + \frac{W}{2} \times 50\right)$$

$$\therefore M_{CD} = +15.915W$$

Now,

$$\sigma_C = \frac{M_{CD}}{AR}\left[1 + \frac{y_C}{m(R - y_C)}\right]$$

$$= +\frac{15.918W}{490.87 \times 50}\left[1 + \frac{(-12.5)}{0.016(50 - 12.5)}\right]$$

$$\therefore \sigma_C = -0.013W \quad \text{(Compression)}$$

and

$$\sigma_D = +\frac{15.918W}{490.87 \times 50}\left[1 + \frac{12.5}{0.016(50 + 12.5)}\right]$$

$$= 0.0088 \ W \ \text{(Tension)}$$

Stresses at Section CD due to horizontal Loads.
We have, moment at any section MN is given by

$$M_{MN} = -M_A + \frac{PR}{2}(1 - \cos\theta)$$

At section CD, $\theta = 0$

$$\therefore M_{CD} = -M_A + \frac{W}{2}R(1 - \cos 0)$$

$$M_{CD} = -M_A = -9.085W$$

$$\therefore \sigma_C = \frac{P}{A} + \frac{M_{CD}}{AR}\left[1 + \frac{y_C}{m(R + y_C)}\right]$$

$$= \frac{W}{2 \times 490.87} + \frac{(-9.085W)}{490.87 \times 50}\left[1 + \frac{(-12.5)}{0.016(50 - 12.5)}\right]$$

$$\therefore \sigma_C = 0.0083 \ W \ \text{(Tensile)}$$

and

$$\sigma_D = \frac{P}{A} + \frac{M_{CD}}{AR}\left[1 + \frac{y_D}{m(R + y_D)}\right]$$

$$= \frac{W}{2 \times 490.87} + \frac{(-9.085W)}{490.87 \times 50}\left[1 + \frac{12.5}{0.016(50 + 12.5)}\right]$$

$$\therefore \sigma_D = -0.00398W \ \text{(Compression)}$$

Resultant stresses are

$$\sigma_C = (-0.013W + 0.00836W) = -0.00464 \text{ W (Compression)}$$
$$\sigma_D = (0.0088W - 0.00398W) = 0.00482 \text{ W (Tension)}$$

In order to limit the tensile stress to 80 N/mm^2 in the ring, the maximum value of the force in the pulls is given by.
$$0.00482 \ W = 80$$

$$\therefore W = 16597.51 \text{ N or } 16.598 \text{ kN}$$

6.16   Exercises

1. Is the following function a stress function?

$$\phi = -\left(\frac{P}{\pi}\right) r\theta \sin\theta$$

If so, find the corresponding stress. What is the problem solved by this function?

2. Investigate what problem of plane stress is solved by the following stress functions

   (a) $\phi = \frac{P}{K} r\theta \sin\theta$
   (b) $\phi = -\frac{P}{\pi} r\theta \sin\theta$

3. Starting from the stress function $\varphi = A \log r + Br^2 \log r + Cr^2 D$, obtain the stress components σ_r and σ_θ in a pipe subjected to internal pressure p_i and external pressure p_o. Obtain the maximum value of σ_θ when $p_o = 0$ and indicate where it occurs.

4. Check whether the following is a stress function
   $\varphi = c\left[r^2(\alpha - \theta) + r^2 - r^2 \cos^2\theta \tan\alpha\right]$ where α is a constant.

5. Starting frsom the stress function $\varphi = C_r + \frac{C_1}{r} - \frac{(3+\mu)}{8} + C_r^2 r^3$, derive expressions for σ_r and σ_θ in case of a rotating disc of inner radius "a" and outer radius "b". Obtain the maximum values of σ_r and σ_θ.

6. Show that the stress function $\varphi = A \log r + Br^2 \log r + Cr^2 + D$ solves the problem of axisymmetric stress distribution. Obtain expressions for σ_r and σ_θ in case of a pipe subjected to internal pressure p_i and external pressure p_o.

7. Explain axisymmetric problems with examples.

8. Derive the general expression for the stress function in the case of axisymmetric stress distribution.

9. Derive the expression for radial and tangential stress in a thick cylinder subjected to internal and external fluid pressure.

10. A curved bar bent into a arc of a circle having internal radius "a" and external radius "b" is subjected to a bending couple M at its end. Determine the stresses σ_r, σ_θ and $\tau_{r\theta}$.

11. For the stress function, $\varphi = Ar^2 \log r$, where A is a constant, compute the stress components σ_r, σ_θ and $\tau_{r\theta}$.

12. A thick cylinder of inner radius 150 mm and outer radius 200 mm is subjected to an internal pressure of 15 MN/m². Determine the radial and hoop stresses in the cylinder at inner and outer surfaces.

13. The internal and external diameters of a thick hollow cylinder are 80 mm and 120 mm respectively. It is subjected to an external pressure of 40 MN/m², when the internal pressure is 120 MN/m². Calculate the circumferential stresses at the external and internal surfaces and determine the radial and circumferential stresses at the mean radius.

14. A thick-wall cylinder is made of steel ($E = 200$ GPa and $v = 0.29$), has an inside diameter of 20 mm, and an outside diameter of 100 mm. The cylinder is subjected to an internal pressure of 300 MPa. Determine the stress components σ_r and σ_θ at $r = a = 10$ mm, $r = 25$ mm and $r = b = 50$ mm.

15. A long closed cylinder has an internal radius of 100 mm and an external radius of 250 mm. It is subjected to an internal pressure of 80 MPa. Determine the maximum radial, circumferential and axial stresses in the cylinder.

16. A solid disc of radius 200 mm is rotating at a speed of 3000 rpm. Determine the radial and hoop stresses in the disc if $v = 0.3$ and $\rho = 8000$ kg/m³. Also determine the stresses in the disc if a hole of 30 mm is bored at the centre of the disc.

17. A disc of 250 mm diameter has a central hole of 50 mm diameter and runs at 4000 rpm. Calculate the hoop stresses. Take $v = 0.25$ and $\rho = 7800$ kg/m³.

18. A turbine rotor 400 mm external diameter and 200 mm internal diameter revolves at 1000 rpm. Find the maximum hoop and radial stresses assuming the rotor to be thin disc. Take the weight of the rotor as 7700 kg/m³ and Poisson's ratio as 0.3.

19. Check Whether the Following is a Stress Function.

$$\phi = \left(Ar^2 + Br^2 + \frac{C}{r^2} + D \right) \cos 2\theta$$

20. Show that $\phi = \left[Ae^{\alpha y} + Be^{-\alpha y} + Cye^{\alpha y} + Dye^{-\alpha y} \right] \sin \alpha \; x$ represents stress function.

21. The curved beam shown in Fig. 6.23 has a circular cross-section 50 mm in diameter. The inside diameter of the curved beam is 40 mm. Determine the stress at B when $P = 20$ kN.

22. A crane hook carries a load $W = 20$ kN as shown in Fig. 6.24. The cross-section mn of the hook is trapezoidal as shown in the figure. Find the total stresses at points m and n. Use the data as given $b_1 = 40$ mm, $b_2 = 10$ mm, $a = 30$ mm and $c = 120$ mm

Fig. 6.23 Curved beam

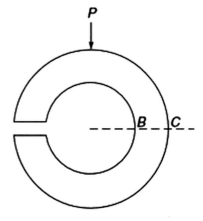

Fig. 6.24 Crane hook

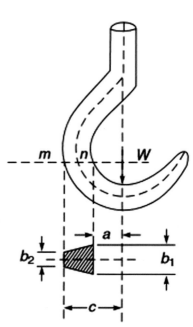

23. A semicircular curved bar is loaded as shown in Fig. 6.25 and has a trapezoidal cross-section. Calculate the tensile stress at point A if $P = 5\,\text{kN}$

24. A curved beam with a circular centreline has a T-section shown in Fig. 6.26. It is subjected to pure bending in its plane of symmetry. The radius of curvature of the concave face is 60 mm. All dimensions of the cross-section are fixed as shown except the thickness t of the stem. Find the proper value of the stem thickness so that the extreme fibre stresses are bending will be numerically equal.

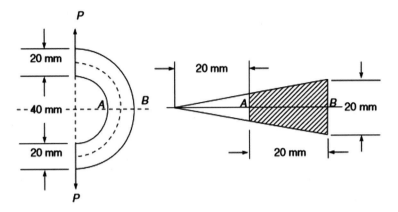

Fig. 6.25 Semicircular curved bar

Fig. 6.26 Curved beam with
T-section

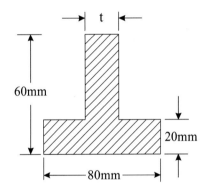

25. A closed ring of mean diameter 200 mm has a rectangular section 50 mm wide by
    a 30 mm thick is loaded as shown in Fig. 6.27. Determine the circumferential
    stress on the inside and outside fibre of the ring at A and B. Assume $E =$
    210 kN/mm²

26. A hook has a triangular cross-section with the dimensions shown in Fig. 6.28.
    The base of the triangle is on the inside of the hook. The load of 20 kN applied
    along a line 50 mm from the inner edge of the shank. Compute the stress at the
    inner and outer fibres.

27. A circular ring of mean radius 40 mm has a circular cross-section with a diameter
    of 25 mm. The ring is subjected to diametrical compressive forces of 30 kN
    along the vertical diameter. Calculate the stresses developed in the vertical
    section under the load and the horizontal section at right angles to the plane of
    loading.

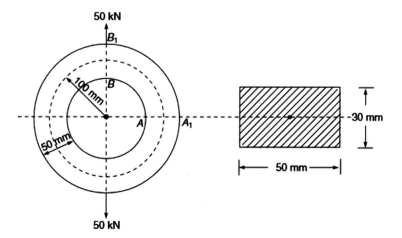

Fig. 6.27 Closed ring

Fig. 6.28 Hook with
triangular section

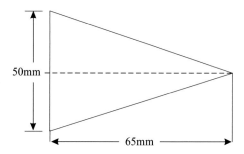

Chapter 7
Torsion of Prismatic Bars

7.1 Introduction

Most of the engineering components are required to withstand torsional or twisting load. Usually, the components that transmit torque, such as propeller shaft and torque tubes of power equipments, are tubular or circular in cross-section. The strength and stiffness of these shafts and torque tubes of uniform circular cross-section can be calculated using simple theory of torsion based on strength of materials. However, slender members with other than circular cross-section are also often used. The strength and stiffness of these members can be determined only by using more sophisticated analyses.

From the study of elementary strength of materials, two important expressions related to the torsion of circular bars were developed. They are

$$\tau = \frac{M_t r}{J} \tag{7.1}$$

and

$$\theta = \frac{1}{L} \int_L \frac{M_t dz}{GJ} \tag{7.2}$$

Here τ represents the shear stress, M_t the applied torque, r the radius at which the stress is required, G the shear modulus or modulus of rigidity, θ the angle of twist per unit longitudinal length, L the length, and z the axial co-ordinate. Also, J is polar moment of inertia which is defined by $\int_A r^2 dA$.

The following are the assumptions associated with the elementary approach in deriving Eqs. (7.1) and (7.2).

© The Author(s), under exclusive license to Springer Nature Singapore Pte Ltd. 2021
T. G. Sitharam and L. Govindaraju, *Theory of Elasticity*,
https://doi.org/10.1007/978-981-33-4650-5_7

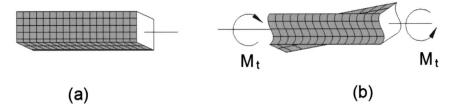

Fig. 7.1 Non-circular sections subjected to torque

1. The material is homogeneous and obeys Hooke's Law.
2. All plane sections perpendicular to the longitudinal axis remain plane following
    the application of a torque, i.e. points in a given cross-sectional plane remain in
    that plane after twisting.
3. Subsequent to twisting, cross-sections are undistorted in their individual planes,
    i.e. the shearing strain varies linearly with the distance from the central axis.
4. Angle of twist per unit length is constant.

While treating non-circular prismatic bars such as shown in Fig. 7.1, initially
plane cross-sections (Fig. 7.1a) experience out-of-plane deformation or "warping"
(Fig. 7.1b), and therefore, in the above assumptions second and third are no longer
appropriate. Consequently, a different analytical approach is employed, which is the
theory of elasticity.

7.2   General Solution of the Torsion Problem

The correct solution of the problem of torsion of bars by couples applied at the ends
was given by Saint–Venant. He used the semi-inverse method. In the beginning, he
made certain assumptions for the deformation of the twisted bar and showed that
these assumptions could satisfy the equations of equilibrium given by

$$\frac{\partial \sigma_x}{\partial x} + \frac{\partial \tau_{xy}}{\partial y} + \frac{\partial \tau_{xz}}{\partial z} + F_x = 0$$

$$\frac{\partial \sigma_y}{\partial y} + \frac{\partial \tau_{xy}}{\partial x} + \frac{\partial \tau_{yz}}{\partial z} + F_y = 0$$

$$\frac{\partial \sigma_z}{\partial z} + \frac{\partial \tau_{xz}}{\partial x} + \frac{\partial \tau_{yz}}{\partial y} + F_z = 0$$

and the boundary conditions such as

$$\overline{X} = \sigma_x l + \tau_{xy} m + \tau_{xz} n$$

$$\overline{Y} = \sigma_y m + \tau_{yz} n + \tau_{xy} l$$

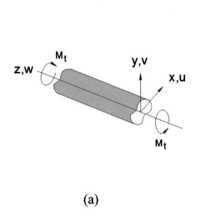

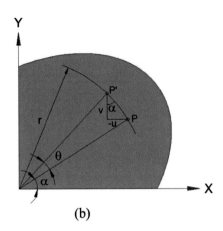

(a) (b)

Fig. 7.2 Bars subjected to torsion

$$\overline{Z} = \sigma_z n + \tau_{xz} l + \tau_{yz} m$$

in which F_x, F_y, F_z are the body forces, $\overline{X}$, $\overline{Y}$, $\overline{Z}$ are the components of the surface forces per unit area and l, m, n are the direction cosines.

Also from the uniqueness of solutions of the elasticity equations, it follows that the torques on the ends are applied as shear stress in exactly the manner required by the solution itself.

Now, consider a prismatic bar of constant arbitrary cross-section subjected to equal and opposite twisting moments applied at the ends, as shown in Fig. 7.2a.

Saint–Venant assumes that the deformation of the twisted shaft consists of

1. Rotations of cross-sections of the shaft as in the case of a circular shaft and
2. Warping of the cross-sections that are the same for all cross-sections.

The origin of X, Y, Z in the figure is located at the centre of the twist of the cross-section, about which the cross-section rotates during twisting. Figure 7.1b shows the partial end view of the bar (and could represent any section). An arbitrary point on the cross-section, point $P(x, y)$, located a distance r from centre of twist A, has moved to $P'(x - u, y + v)$ as a result of torsion. Assuming that no rotation occurs at end $z = 0$ and that θ is small, the x and y displacements of P are respectively:

$$u = -(r\theta_z) \sin \alpha$$

But

$$\sin \alpha = y/r$$

Therefore,

$$u = -(r\theta_z)y/r = -y\theta_z \tag{a}$$

Similarly,

$$v = (r\theta_z)\cos\alpha = (r\theta_z)x/r = x\theta_z \tag{b}$$

where θ_z is the angle of rotation of the cross-section at a distance z from the origin. The warping of cross-sections is defined by a function ψ as

$$w = \theta\psi(x, y) \tag{c}$$

Here, the equations (a) and (b) specify the rigid body rotation of any cross-section through a small angle θ_z. However, with the assumed displacements (a), (b) and (c), we calculate the components of strain from the equations given below.

$$\varepsilon_x = \frac{\partial u}{\partial x}, \varepsilon_y = \frac{\partial v}{\partial y}, \varepsilon_z = \frac{\partial w}{\partial z}$$

$$\gamma_{xx} = \frac{\partial u}{\partial y} + \frac{\partial v}{\partial x}, \gamma_{yz} = \frac{\partial v}{\partial z} + \frac{\partial w}{\partial y}$$

and

$$\gamma_{zx} = \frac{\partial w}{\partial x} + \frac{\partial u}{\partial z}$$

Substituting (a), (b) and (c) in the above equations, we obtain

$$\varepsilon_x = \varepsilon_y = \varepsilon_z = \gamma_{xy} = 0$$

$$\gamma_{xz} = \frac{\partial w}{\partial x} - y\theta = \left(\theta\frac{\partial\psi}{\partial x} - y\theta\right)$$

or

$$\gamma_{xz} = \theta\left(\frac{\partial\psi}{\partial x} - y\right)$$

and

$$\gamma_{yz} = \frac{\partial w}{\partial y} + x\theta = \left(\theta\frac{\partial\psi}{\partial y} + x\theta\right)$$

or

$$\gamma_{yz} = \theta\left(\frac{\partial\psi}{\partial y} + x\right)$$

Also, by Hooke's law, the stress–strain relationships are given by

$$\sigma_x = 2G\varepsilon_x + \lambda e, \quad \tau_{xy} = G\gamma_{xy}$$
$$\sigma_y = 2G\varepsilon_y + \lambda e, \quad \tau_{yz} = G\gamma_{yz}$$
$$\sigma_z = 2G\varepsilon_2 + \lambda e, \quad \tau_{xz} = G\gamma_{xz}$$

where $e = \varepsilon_x + \varepsilon_y + \varepsilon_z$.

and

$$\lambda = \frac{v\,E}{(1+v)(1-2v)}.$$

Substituting (a), (b) and (c) in the above equations, we obtain

$$\sigma_x = \sigma_y = \sigma_z = \tau_{xy} = 0$$

$$\tau_{xz} = G\left(\frac{\partial w}{\partial x} - y\theta\right) = G\theta\left(\frac{\partial\psi}{\partial x} - y\right) \tag{d}$$

$$\tau_{yz} = G\left(\frac{\partial w}{\partial y} + x\theta\right) = G\theta\left(\frac{\partial\psi}{\partial y} + x\right) \tag{e}$$

It can be observed that with the assumptions (a), (b) and (c) regarding deformation, there will be no normal stresses acting between the longitudinal fibres of the shaft or in the longitudinal direction of those fibres. Also, there will be no distortion in the planes of cross-sections, since ε_x, ε_y and γ_{xy} vanish. We have at each point, pure shear defined by the components τ_{xz} and τ_{yz}.

However, the stress components should satisfy the equations of equilibrium given by:

$$\frac{\partial\sigma_x}{\partial x} + \frac{\partial\tau_{xy}}{\partial y} + \frac{\partial\tau_{xz}}{\partial z} + F_x = 0$$

$$\frac{\partial\sigma_y}{\partial y} + \frac{\partial\tau_{xy}}{\partial x} + \frac{\partial\tau_{yz}}{\partial z} + F_y = 0$$

$$\frac{\partial\sigma_z}{\partial z} + \frac{\partial\tau_{xz}}{\partial x} + \frac{\partial\tau_{yz}}{\partial y} + F_z = 0$$

Assuming negligible body forces, and substituting the stress components into equilibrium equations, we obtain

$$\frac{\partial\tau_{xz}}{\partial z} = 0, \quad \frac{\partial\tau_{zy}}{\partial z} = 0, \quad \frac{\partial\tau_{xz}}{\partial x} + \frac{\partial\tau_{zy}}{\partial y} = 0 \tag{7.3}$$

Also, the function $\psi(x, y)$, defining warping of cross-section must be determined by the equations of equilibrium.

Therefore, we find that the function ψ must satisfy the equation

$$\frac{\partial^2 \psi}{\partial x^2} + \frac{\partial^2 \psi}{\partial y^2} = 0 \tag{7.3a}$$

Now, differentiating equation (d) with respect to y and the equation (e) with respect to x, and subtracting we get an equation of compatibility.

Hence,

$$\frac{\partial \tau_{xz}}{\partial y} = -G\theta$$

$$\frac{\partial \tau_{yz}}{\partial x} = G\theta$$

$$\frac{\partial \tau_{xz}}{\partial y} - \frac{\partial \tau_{yz}}{\partial x} = -G\theta - G\theta$$

$$= -2G\theta \tag{7.4}$$

Therefore, the stress in a bar of arbitrary section may be determined by solving Eqs. (7.3) and (7.4) along with the given boundary conditions.

7.3 Boundary Conditions

Now, consider the boundary conditions given by

$$\overline{X} = \sigma_x l + \tau_{xy} m + \tau_{xz} n$$

$$\overline{Y} = \sigma_y m + \tau_{yz} n + \tau_{xy} l$$

$$\overline{Z} = \sigma_z n + \tau_{xz} l + \tau_{yz} m$$

For the lateral surface of the bar, which is free from external forces acting on the boundary and the normal n to the surface is perpendicular to the z-axis. The first two equations are identically satisfied and the third gives,

$$\tau_{xz} l + \tau_{yz} m = 0 \tag{7.5}$$

which means that the resultant shearing stress at the boundary is directed along the tangent to the boundary, as shown in Fig. 7.3.

Considering an infinitesimal element abc at the boundary and assuming that S is increasing in the direction from c to a,

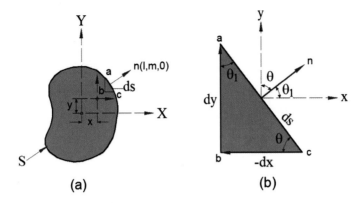

Fig. 7.3 Cross-section of the bar and boundary conditions

$$\ell = \cos(n, x) = \frac{dy}{dS}$$

$$m = \cos(n, y) = -\frac{dx}{dS}$$

Hence, Eq. (7.5) becomes

$$\tau_{xz}\left(\frac{dy}{dS}\right) - \tau_{yz}\left(\frac{dx}{dS}\right) = 0$$

$$\text{or} \quad \left(\frac{\partial \psi}{\partial x} - y\right)\left(\frac{dy}{dS}\right) - \left(\frac{\partial \psi}{\partial y} + x\right)\left(\frac{dx}{dS}\right) = 0 \qquad (7.6)$$

Thus, each problem of torsion is reduced to the problem of finding a function ψ satisfying Eq. (7.3a) and the boundary condition (7.6).

7.4 Stress Function Method

Similar to the case of beams, the torsion problem formulated above is commonly solved by introducing a single stress function. This procedure has the advantage of leading to simpler boundary conditions as compared to Eq. (7.6). The method was proposed by Prandtl (1903). In this method, the principal unknowns are the stress components rather than the displacement components as in the previous approach.

Based on the result of the torsion of the circular shaft, let the non-vanishing components be τ_{zx} and τ_{yz}. The remaining stress components σ_x, σ_y and σ_z and τ_{xy} are assumed to be zero. In order to satisfy the equations of equilibrium, we should have

$$\frac{\partial \tau_{xz}}{\partial z} = 0, \quad \frac{\partial \tau_{yz}}{\partial z} = 0, \quad \frac{\partial \tau_{xz}}{\partial x} + \frac{\partial \tau_{yz}}{\partial y} = 0.$$

The first two are already satisfied since τ_{xz} and τ_{yz}, as given by Equations (d) and (e) are independent of z.

In order to satisfy the third condition, we assume a function $\phi(x, y)$ called Prandtl stress function such that

$$\tau_{xz} = \frac{\partial \phi}{\partial y}, \quad \tau_{yz} = -\frac{\partial \phi}{\partial x} \tag{7.7}$$

With this stress function (called Prandtl torsion stress function), the third condition is also satisfied. The assumed stress components, if they are to be proper elasticity solutions, have to satisfy the compatibility conditions. We can substitute these directly into the stress equations of compatibility. Alternately, we can determine the strains corresponding to the assumed stresses and then apply the strain compatibility conditions.

Therefore from Eqs. (7.7), (d) and (e), we have

$$\frac{\partial \phi}{\partial y} = G\theta \left(\frac{\partial \psi}{\partial x} - y \right) \quad -\frac{\partial \phi}{\partial x} = G\theta \left(\frac{\partial \psi}{\partial y} + x \right)$$

Eliminating ψ by differentiating the first with respect to y, the second with respect to x, and subtracting from the first, we find that the stress function must satisfy the differential equation

$$\frac{\partial^2 \phi}{\partial x^2} + \frac{\partial^2 \phi}{\partial y^2} = -2G\theta$$

$$\text{or} \quad \frac{\partial^2 \phi}{\partial x^2} + \frac{\partial^2 \phi}{\partial y^2} = 2G\theta \tag{7.8}$$

The boundary condition (7.5) becomes, introducing Eq. (7.7)

$$\frac{\partial \phi}{\partial y} \frac{dy}{dS} + \frac{\partial \phi}{\partial x} \frac{dx}{dS} = \frac{d\phi}{dS} = 0 \tag{7.9}$$

This shows that the stress function ϕ must be constant along the boundary of the cross-section. In the case of singly connected sections, example, for solid bars, this constant can be arbitrarily chosen. Since the stress components depend only on the differentials of ϕ, for a simply connected region, no loss of generality is involved in assuming $\phi = 0$ on S. However, for a multi-connected region, example shaft having holes, certain additional conditions of compatibility are imposed. Thus, the determination of stress distribution over a cross-section of a twisted bar is used in finding the function ϕ that satisfies.

Equation (7.8) and is zero at the boundary.

Conditions at the Ends of the Twisted Bar

At the ends of the twisted bar in which the normals are parallel to the z-axis, $\cos(n, z)$ $= n = \pm 1, l = m = 0$. Therefore, for $T_z = 0$. Equations (2.22a), (2.22b) and (2.22c) become

$$T_x = \pm \tau_{xz} \text{ and } T_y = \pm \tau_{yz} \tag{7.10}$$

on the two ends faces, the summation of forces over the ends of the bar in x and y directions should be zero

i.e.,

$$\iint T_x dx\, dy = \int \tau_{xz} dx\, dy = \int \frac{d\phi}{dy} dx\, dy$$

$$= \int dx \int_{y_1}^{y_2} \frac{d\phi}{dy} dy$$

$$= \int [\phi]_{y_1}^{y_2} dx \tag{7.11}$$

Here y_1, and y_2 represent the y co-ordinates of points on the surface. Further, as $\phi = $ constant on the surface of the bar, values of ϕ corresponding to y_1, and y_2 and must to equal to a constant and hence $\phi_1 = \phi_2 = $ constant.

$$\therefore \iint T_x dx\, dy = \int [\phi]_{y_1}^{y_2} dx = \int (\phi_2 - \phi_1) dx = 0 \tag{7.12}$$

Similarly, considering the resultant in y-direction, it can be shown that

$$\iint T_x dx\, dy = \iint \tau_{yz} dx\, dy = 0 \tag{7.13}$$

Thus, the resultant of the forces distributed over the ends of the bar is zero, and these forces represent a couple the magnitude of which is

$$M_t = \iint \left(x\, \tau_{yz} - y\, \tau_{xs} \right) dx dy$$

$$= -\iint x \frac{\partial \phi}{\partial x} dx dy - \iint y \frac{\partial \phi}{\partial y} dx dy$$

Solving,

$$M_t = -\int x[\phi]_{x_1}^{x_2} dy + \iint \phi dx dy - \int y[\phi]_{y_1}^{y_2} dx + \iint \phi dx dy \tag{7.14}$$

As ϕ = constant at the boundary and x_1, x_2, y_1 and y_2 represent points on the lateral surface, then

$$M_t = \iint \phi \, dx \, dy + \iint \phi \, dx \, dy = 2 \iint \phi \, dx \, dy \qquad (7.15)$$

Hence, it is observed that each of the integrals in Eq. (7.15) contributing one half of the torque due to τ_{xz} and the other half due to τ_{yz}.

Thus all the differential equations and boundary conditions are satisfied if the stress function ϕ, obeys Eq. (7.8) and is used to obtain M_t, and the solution obtained in this manner is the exact solution of the torsion problem.

7.5 Torsion of Circular Cross-Section

The Laplace equation is given by

$$\frac{\partial^2 \psi}{\partial x^2} + \frac{\partial^2 \psi}{\partial y^2} = 0$$

where ψ = warping function.

The simplest solution to the above equation is

$$\psi = \text{constant} = C$$

But the boundary condition is given by Eq. (7.6) is

$$\left(\frac{\partial \psi}{\partial x} - y \right) \left(\frac{dy}{dS} \right) - \left(\frac{\partial \psi}{\partial y} + x \right) \left(\frac{dx}{dS} \right) = 0$$

Therefore, with $\psi = C$, the above boundary condition becomes

$$(0 - y)(dy/dS) - (0 + x)(dx/dS) = 0$$

$$-y \frac{dy}{dS} - x \frac{dx}{dS} = 0$$

or

$$\frac{d}{dS} \frac{x^2 + y^2}{2} = 0$$

i.e.

$$x^2 + y^2 = \text{constant}$$

where (x, y) are the co-ordinates of any point on the boundary. Hence, the boundary is a circle.

From Equation (c), we can write

$$w = \theta \psi (x, y)$$

i.e.

$$w = \theta C.$$

The polar moment of inertia for the section is

$$J = \int \int (x^2 + y^2) dxdy = I_P$$

But

$$M_t = G I_P \theta$$

or

$$\theta = \frac{M_t}{G I_P}.$$

Therefore,

$$w = \frac{M_t C}{G I_P}$$

which is a constant. Since the fixed end has zero w at least at one point, w is zero at every cross-section (other than the rigid body displacement). Thus the cross-section does not warp.

Further, the shear stresses are given by the Equations (d) and (e) as

$$\tau_{xz} = G \left(\frac{\partial w}{\partial x} - y\theta \right) = G\theta \left(\frac{\partial \psi}{\partial x} - y \right)$$

$$\tau_{yz} = G \left(\frac{\partial w}{\partial y} + x\theta \right) = G\theta \left(\frac{\partial \psi}{\partial y} + x \right)$$

$$\therefore \tau_{xz} = -G\theta y$$

and

$$\tau_{yz} = G\theta x$$

or

$$\tau_{xz} = -\frac{M_t x}{I_P}$$

and

$$\tau_{yz} = G\theta x$$
$$= G\frac{M_t}{GI_P x}$$

hence, $\tau_{yz} = \frac{M_t x}{I_P}$.

Therefore, the direction of the resultant shear stress τ is such that, from Fig. 7.4

$$\tan \alpha = \frac{\tau_{yz}}{\tau_{xz}} = \frac{M_t x/I_P}{-M_t y/I_P} = -x/y$$

Hence, the resultant shear stress is perpendicular to the radius.
Further,

$$\tau^2 = \tau_{yz}^2 + \tau_{xz}^2$$
$$\tau^2 = M_t^2(x^2 + y^2)/I_p^2$$

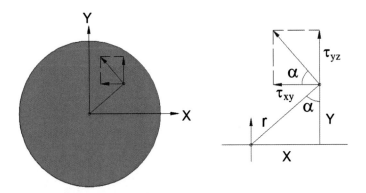

Fig. 7.4 Circular bar under torsion

or

$$\tau = \frac{M_{\mathrm{t}}}{I_{\mathrm{P}}}\sqrt{x^2 + y^2}$$

Therefore,

$$\tau = \frac{M_{\mathrm{t}} . r}{I_{\mathrm{P}}}$$

or

$$\tau = \frac{M_{\mathrm{t}} . r}{J}(\text{since } J = I_{\mathrm{P}})$$

where r is the radial distance of the point (x, y). Hence, all the results of the elementary analysis are justified.

7.6 Torsion of Elliptical Cross-Section

Let the warping function is given by

$$\psi = Axy \qquad\qquad (7.16)$$

where A is a constant. This also satisfies the Laplace equation. The boundary condition gives

$$(Ay - y)\frac{\mathrm{d}y}{\mathrm{d}S} - (Ax + x)\frac{\mathrm{d}x}{\mathrm{d}S} = 0$$

or

$$y(A - 1)\frac{\mathrm{d}y}{\mathrm{d}S} - x(A + 1)\frac{\mathrm{d}x}{\mathrm{d}S} = 0$$

i.e.

$$(A + 1)2x\frac{\mathrm{d}x}{\mathrm{d}S} - (A - 1)2y\frac{\mathrm{d}y}{\mathrm{d}S} = 0$$

or

$$\frac{d}{dS}[(A+1)x^2 - (A-1)y^2] = 0.$$

Integrating, we get

$$(1+A)x^2 + (1-A)y^2 = \text{constant.}$$

This is of the form

$$\frac{x^2}{a^2} + \frac{y^2}{b^2} = 1$$

These two are identical if

$$\frac{a^2}{b^2} = \frac{1-A}{1+A}$$

or

$$A = \frac{b^2 - a^2}{b^2 + a^2}$$

Therefore, the function given by

$$\psi = \frac{b^2 - a^2}{b^2 + a^2}xy \tag{7.17}$$

represents the warping function for an elliptic cylinder with semi-axes a and b under torsion. The value of polar moment of inertia J is

$$J = \int\int (x^2 + y^2 + Ax^2 - Ay^2)dxdy$$

$$= (A+1)\int\int x^2dxdy + (1-A)\int\int y^2dxdy \tag{7.18}$$

$$J = (A+1)I_y + (1-A)I_x \tag{7.19}$$

where $I_x = \frac{\pi ab^3}{4}$ and $I_y = \frac{\pi a^3 b}{4}$.

Substituting the above values in (7.19), we obtain

$$J = \frac{\pi a^3 b^3}{a^2 + b^2}$$

But

$$\theta = \frac{M_t}{GI_p} = \frac{M_t}{GJ}$$

Therefore,

$$M_t = GJ\theta$$

$$= G\theta \frac{\pi a^3 b^3}{a^2 + b^2}$$

or

$\theta = \frac{M_t}{G} \frac{a^2 + b^2}{\pi a^3 b^3}$

The shearing stresses are given by

$$\tau_{yz} = G\theta\left(\frac{\partial \psi}{\partial y} + x\right)$$

$$= M_t \frac{a^2 + b^2}{\pi a^3 b^3}\left(\frac{b^2 - a^2}{b^2 + a^2} + 1\right)x$$

or

$$\tau_{yz} = \frac{2M_t x}{\pi a^3 b}$$

Similarly,

$$\tau_{xz} = \frac{2M_t y}{\pi a b^3}$$

Therefore, the resultant shearing stress at any point (x, y) is

$$\tau = \sqrt{\tau_{yz}^2 + \tau_{xz}^2} = \frac{2M_t}{\pi a^3 b^3}\left[b^4 x^2 + a^4 y^2\right]^{\frac{1}{2}} \tag{7.20}$$

Determination of Maximum Shear Stress

To determine where the maximum shear stress occurs, substitute for x^2 from

$$\frac{x^2}{a^2} + \frac{y^2}{b^2} = 1,$$

or

$$x^2 = a^2\left(1 - y^2/b^2\right)$$

and

Fig. 7.5 Cross-section of
elliptic bar and contour lines
of w

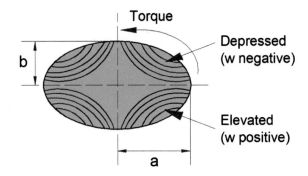

$$\tau = \frac{2M_t}{\pi a^3 b^3}\left[a^2 b^4 + a^2(a^2 - b^2)y^2\right]^{\frac{1}{2}}$$

Since all terms under the radical (power 1/2) are positive, the maximum shear stress occurs when y is maximum, i.e. when $y = b$. Thus, maximum shear stress τ_{max} occurs at the ends of the minor axis and its value is

$$\tau_{max} = \frac{2M_t}{\pi a^3 b^3}(a^4 b^2)^{1/2}$$

Therefore,

$\tau_{max} = \frac{2M_t}{\pi a b^2}.$

For $a = b$, this formula coincides with the well-known formula for circular cross-section. Knowing the warping function, the displacement w can be easily determined.

Therefore,

$$w = \theta\psi = \frac{M_t(b^2 - a^2)}{\pi a^3 b^3 G}xy \qquad (7.21)$$

The contour lines giving $w = $ constant are the hyperbolas shown in Fig. 7.5 having the principal axes of the ellipse as asymptotes. ·

7.7   Prandtl's Membrane Analogy

It becomes evident that for bars with more complicated cross-sectional shapes, more analytical solutions are involved and hence become difficult. In such situations, it is desirable to use other techniques—experimental or otherwise. The membrane analogy introduced by Prandtl is one of the useful solutions in such cases.

Consider a thin homogeneous membrane, like a thin rubber sheet be stretched with uniform tension fixed at its edge which is a given curve (the cross-section of the shaft) in the x y-plane as shown in Fig. 7.6.

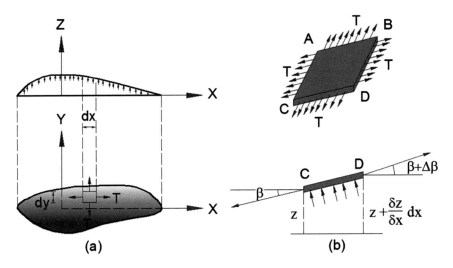

Fig. 7.6 Stretching of a membrane

When the membrane is subjected to a constant lateral pressure p, it undergoes a small displacement z where z is a function of x and y.

Consider the equilibrium of an element ABCD of the membrane after deformation. Let T be the uniform tension per unit length of the membrane. The value of the initial tension T is large enough to ignore its change when the membrane is blown up by the small pressure p. On the face AC, the force acting is T.dy. This is inclined at an angle β to the x-axis. Also, tan β is the slope of the face AB and is equal to $\frac{\partial z}{\partial x}$. Hence, the component of Tdy in z-direction is $\left(-T dy \frac{\partial z}{\partial x}\right)$. The force on face BD is also Fdy but is inclined at an angle $(\beta + \Delta\beta)$ to the x-axis. Its slope is, therefore,

$$\left[\frac{\partial z}{\partial x} + \frac{\partial}{\partial x}\left(\frac{\partial z}{\partial x} \right) dx \right]$$

and the component of the force in the z-direction is

$$T dy \left[\frac{\partial z}{\partial x} + \frac{\partial}{\partial x}\left(\frac{\partial z}{\partial x} \right) dx \right]$$

Similarly, the components of the forces $T dx$ acting on face AB and CD are

$$-T dx \frac{\partial z}{\partial y} \quad \text{and} \quad T dx \left[\frac{\partial z}{\partial y} + \frac{\partial}{\partial y}\left(\frac{\partial z}{\partial y} \right) dy \right]$$

Therefore, the resultant force in z-direction due to tension T

$$= -T\,dy\frac{\partial z}{\partial x} + T\,dy\left[\frac{\partial z}{\partial x} + \frac{\partial^2 z}{\partial x^2}dx\right]$$

$$- T\,dx\frac{\partial z}{\partial y} + T\,dx\left[\frac{\partial z}{\partial y} + \frac{\partial^2 z}{\partial y^2}dy\right]$$

$$= T\left(\frac{\partial^2 z}{\partial x^2} + \frac{\partial^2 z}{\partial y^2}\right)dx\,dy$$

But the force p acting upwards on the membrane element ABCD is $p\,dxdy$, assuming that the membrane deflection is small.

Hence, for equilibrium,

$$T\left(\frac{\partial^2 z}{\partial x^2} + \frac{\partial^2 z}{\partial y^2}\right) = -p$$

or

$$\left(\frac{\partial^2 z}{\partial x^2} + \frac{\partial^2 z}{\partial y^2}\right) = -p/T \tag{7.22}$$

Now, if the membrane tension T or the air pressure p is adjusted in such a way that p/T becomes numerically equal to $2G\theta$, then Eq. (7.22) of the membrane becomes identical to Eq. (7.8) of the torsion stress function ϕ. Also, if the membrane height z remains zero at the boundary contour of the section, then the height z of the membrane becomes numerically equal to the torsion stress function $\phi = 0$. The slopes of the membrane are then equal to the shear stresses and these are in a direction perpendicular to that of the slope.

Further, the twisting moment is numerically equivalent to twice the volume under the membrane [Eq. (7.15)] (Table 7.1).

The membrane analogy provides a useful experimental technique. It also serves as the basis for obtaining approximate analytical solutions for bars of narrow cross-section as well as for member of open thin-walled section.

Table 7.1 Analogy between torsion and membrane problems

Membrane problem	Torsion problem
z	ϕ
$\frac{1}{S}$	G
p	2θ
$-\frac{\partial z}{\partial x}, \frac{\partial z}{\partial y}$	τ_{zy}, τ_{zx}
2 (volume beneath membrane)	M_t

7.8   Torsion of Thin-Walled Sections

Consider a thin-walled tube subjected to torsion. The thickness of the tube may not be uniform as shown in Fig. 7.7.

As the thickness is small and the boundaries are free, the shear stresses on the boundary are parallel. Let τ be the magnitude of shear stress and t is the thickness of the tube.

Now, consider the equilibrium of an element of length dl as shown in Fig. 7.7. The areas of cut faces AB and CD are $t_1 \, dl$ and $t_2 \, dl$ respectively. The shear stresses (complementary shears) are τ_1 and τ_2.

For equilibrium in z-direction, we have

$$-\tau_1 t_1 dl + \tau_2 t_2 dl = 0$$

Therefore, $\tau_1 \, t_1 = \tau_2 \, t_2 = q = \text{constant}$.

Hence, the quantity $\tau \, t$ is constant. This is called the shear flow q.

Determination of Torque Due to Shear and Rotation

Consider the torque of the shear about point O (Fig. 7.8). The force acting on the elementary length dS of the tube is given by

$$\Delta F = \tau t \, dS = q \, dS$$

The moment arm about point O is h and hence the torque

$$\Delta M_t = (q \, dS) h$$

Therefore,

$$\Delta M_t = 2q \, dA$$

where dA is the area of the triangle enclosed at point O by the base dS.

Hence the total torque is

$$M_t = \Sigma 2q \, dA$$

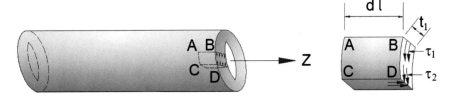

Fig. 7.7 Torsion of thin-walled sections

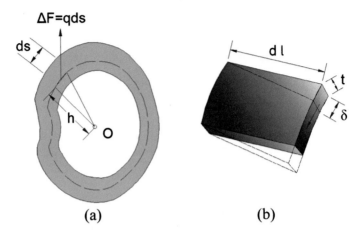

Fig. 7.8 Cross-section of a thin-walled tube and torque due to shear

Therefore,

$$M_t = 2q\,A \qquad (7.23)$$

where A is the area enclosed by the centre line of the tube. Equation (7.23) is generally known as the "Bredt-Batho" formula.

Determination of the Twist of the Tube

In order to determine the twist of the tube, Castigliano's theorem is used. Referring to Fig. 7.8b, the shear force on the element is $\tau\,t\,dS = q\,dS$. Due to shear strain γ, the force does work equal to ΔU

i.e.

$$
\begin{aligned}
\Delta U &= \frac{1}{2}(\tau t\,dS)\delta \\
&= \frac{1}{2}(\tau t\,dS)\gamma \cdot dl \\
&= \frac{1}{2}(\tau t\,dS)\cdot dl\cdot\frac{\tau}{G}\ (\text{since } \tau = G\gamma) \\
&= \frac{\tau^2 t^2\,dS\,dl}{2Gt} \\
&= \frac{q^2\,dS\,dl}{2Gt} \\
\Delta U &= \frac{M_t^2\,dl}{8A^2 G}\cdot\frac{dS}{t}
\end{aligned}
$$

Therefore, the total elastic strain energy is

$$U = \frac{M_t^2 dl}{8A^2 G} \oint \frac{dS}{t}$$

Hence, the twist or the rotation per unit length ($dl = 1$) is

$$\theta = \frac{\partial U}{\partial M_t} = \frac{M_t}{4A^2 G} \oint \frac{dS}{t}$$

or

$$\theta = \frac{2qA}{4A^2 G} \oint \frac{dS}{t}$$

or

$$\theta = \frac{q}{2AG} \oint \frac{dS}{t} \tag{7.24}$$

7.9 Torsion of Thin-Walled Multiple-Cell Closed Sections

Consider a two-cell section shown in Fig. 7.9. Let A_1 and A_2 be the areas of the cells 1 and 2, respectively. Consider the equilibrium of an element at the junction as shown in Fig. 7.9b. In the direction of the axis of the tube, we can write

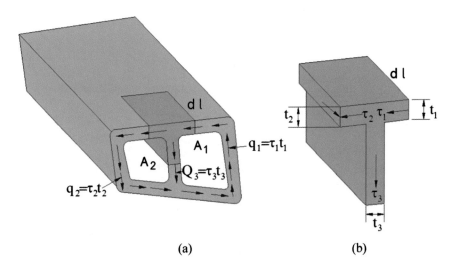

(a)   (b)

Fig. 7.9 Torsion of thin-walled multiple-cell closed section

Fig. 7.10 Cross-section of a
thin-walled multiple-cell
beam and moment axis

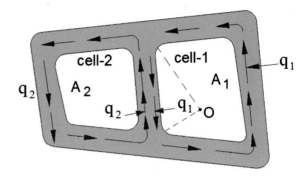

$$-\tau_1 t_1 dl + \tau_2 t_2 dl + \tau_3 t_3 dl = 0$$

or

$$\tau_1 t_1 = \tau_2 t_2 + \tau_3 t_3$$

i.e.

$$q_1 = q_2 + q_3 \tag{7.25}$$

This is again equivalent to a fluid flow dividing itself into two streams. Now, choose moment axis, such as point O as shown in Fig. 7.10.

The shear flow in the web is considered to be made of q_1 and $-q_2$, since $q_3 = q_1 - q_2$.

Moment about point O due to q_1 flowing in cell 1 (including web) is

$$M_{t1} = 2q_1 A_1 \tag{7.26}$$

Similarly, the moment about point O due to q_2 flowing in cell 2 (including web) is

$$M_{t2} = 2q_2(A_2 + A_1) - 2q_2 A_1 \tag{7.27}$$

The second term with the negative sign on the right-hand side is the moment due to shear flow q_2 in the middle web.

Therefore, the total torque is

$$M_t = M_{t_1} + M_{t_2}$$
$$M_t = 2q_1 A_1 + 2q_2 A_2 \tag{7.28}$$

To Find the Twist (θ)
For continuity, the twist of each cell should be the same.
    We have

$$\theta = \frac{q}{2AG} \oint \frac{dS}{t}$$

or

$$G\theta = \frac{1}{A} \int \frac{qdS}{t}$$

    Let

$a_1 = \oint \frac{dS}{t}$ for Cell 1 including the web
$a_2 = \oint \frac{dS}{t}$ for Cell 2 including the web
$a_{12} = \oint \frac{dS}{t}$ for the web only

    Then for Cell 1

$$2G\theta = \frac{1}{A_1}(a_1q_1 - a_{12}q_2) \tag{7.29}$$

    For Cell 2

$$2G\theta = \frac{1}{A_2}(a_2q_2 - a_{12}q_1) \tag{7.30}$$

    Equations (7.28), (7.29) and (7.30) are sufficient to solve for q_1, q_2 and θ.

7.10   Numerical Examples

Example 7.1 A hollow aluminium tube of rectangular cross-section shown in Fig. 7.11 is subjected to a torque of 56,500 mN along its longitudinal axis. Determine the shearing stresses and the angle of twist. Assume $G = 27.6 \times 10^9$ N/m^2.

Solution The above figure shows the membrane surface $ABCD$

$$\text{Now, the applied torque} = M_t = 2qA$$
$$56{,}500 = 29(0.5 \times 0.25)$$
$$56{,}500 = 0.25q$$

hence, $q = 226{,}000$ N/m.
    Now, the shearing stresses are

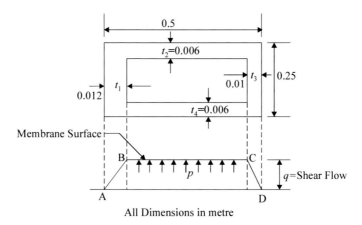

Fig. 7.11 Hollow aluminium tube

$$\tau_1 = \frac{q}{t_1} = \frac{226,000}{0.012} = 18.833 \times 10^6 \, \text{N/m}^2$$

$$\tau_2 = \frac{q}{t_2} = \frac{226,000}{0.006} = 37.667 \times 10^6 \, \text{N/m}^2$$

$$\tau_3 = \frac{226,000}{0.01} = 22.6 \times 10^6 \, \text{N/m}^2$$

Now, the angle of twist per unit length is

$$\theta = \frac{q}{2GA} \oint \frac{ds}{t}$$

Therefore,

$$\theta = \frac{226,000}{2 \times 27.6 \times 10^9 \times 0.125} \left[\frac{0.25}{0.012} + \frac{0.5}{0.006}(2) + \frac{0.25}{0.01} \right]$$

or

$$\theta = 0.00696014 \, \text{rad/m}.$$

Example 7.2 Figure 7.12 shows a two-cell tubular section as formed by a conventional airfoil shape, and having one interior web. An external torque of 10,000 Nm is acting in a clockwise direction. Determine the internal shear flow distribution. The areas of cell-1 and cell-2 are 680 cm^2 and 2000 cm^2 respectively. The peripheral lengths are indicated in Fig. 7.12.

Solution For Cell 1,

Fig. 7.12 Two-cell tubular section

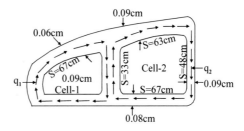

$$a_1 = \oint \frac{dS}{t} \text{(including the web)}$$
$$= \frac{67}{0.06} + \frac{33}{0.09}$$

Therefore,

$$a_1 = 148.3$$

For Cell 2,

$$a_2 = \frac{33}{0.09} + \frac{63}{0.09} + \frac{48}{0.09} + \frac{67}{0.08}$$

Therefore,

$$a_2 = 2409$$

For web,

$$a_{12} = \frac{33}{0.09} = 366$$

Now, for Cell 1,

$$2G\theta = \frac{1}{A_1}(a_1 q_1 - a_{12} q_2)$$
$$= \frac{1}{680}(1483 q_1 - 366 q_2)$$

Therefore,

$$2G\theta = 2.189 q_1 - 0.54 q_2 \qquad \text{(i)}$$

For Cell 2,

$$2G\theta = \frac{1}{A_2}(a_2q_2 - a_{12}q_1)$$

$$= \frac{1}{2000}(2409q_2 - 366q_1)$$

Therefore,

$$2G\theta = 1.20q_2 - 0.18q_1 \qquad\qquad \text{(ii)}$$

Equating (i) and (ii), we get

$$2.18q_1 - 0.54q_2 = 1.20q_2 - 0.18q_1$$

or

$$2.36q_1 - 1.74q_2 = 0$$

or

$$q_2 = 1.36q_1$$

The torque due to shear flows should be equal to the applied torque
Hence, from Eq. (7.28),

$$M_t = 2q_1A_1 + 2q_2A_2$$
$$10{,}000 \times 100 = 2q_1 \times 680 + 2q_2 \times 2000$$
$$= 1360q_1 + 4000q_2$$

Substituting for q_2, we get

$$10{,}000 \times 100 = 1360q_1 + 4000 \times 1.36q_1$$

Therefore,

$$q_1 = 147N \text{ and } q_2 = 200N$$

Example 7.3 A thin-walled steel section shown in Fig. 7.13 is subjected to a twisting moment T. Calculate the shear stresses in the walls and the angle of twist per unit length of the box.

Solution Let A_1 and A_2 be the areas of the cells (1) and (2) respectively.

$$\therefore A_1 = \frac{\pi a^2}{2}$$

Fig. 7.13 Thin-walled steel section

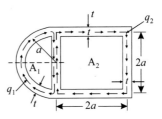

$$A_2 = (2a \times 2a) = 4a^2$$

For Cell (1),

$$a_1 = \oint \frac{ds}{t} \text{(Including the web)}$$

$$a_1 = \left(\frac{\pi a + 2a}{t}\right)$$

For Cell (2),

$$a_2 = \oint \frac{ds}{t}$$

$$= \frac{2a}{t} + \frac{2a}{t} + \frac{2a}{t} + \frac{2a}{t}$$

$$\therefore a_2 = \left(\frac{8a}{t}\right)$$

For web,

$$a_{12} = \left(\frac{2a}{t}\right)$$

Now,
For Cell (1),

$$2G\theta = \frac{1}{A_1}(a_1 q_1 - a_{12} q_2)$$

$$= \frac{2}{\pi a^2}\left[\frac{(\pi a + 2a)}{t}q_1 - \left(\frac{2a}{t}\right)q_2\right]$$

$$= \frac{2a}{\pi t a^2}[(2 + \pi)q_1 - 2q_2]$$

$$\therefore 2G\theta = \frac{2}{\pi a t}[(\pi + 2)q_1 - 2q_2] \qquad \text{(a)}$$

For Cell (2),

$$2G\theta = \frac{1}{A_2}(a_2 q_2 - a_{12} q_1)$$

$$= \frac{1}{4a^2}\left[\frac{8a}{t}q_2 - \frac{2a}{t}q_1\right]$$

$$= \frac{2a}{4a^2 t}[4q_2 - q_1]$$

$$\therefore 2G\theta = \frac{1}{2at}[4q_2 - q_1] \qquad\qquad\text{(b)}$$

Equating (1) and (2), we get,

$$\frac{2}{\pi at}[(\pi + 2)q_1 - 2q_2] = \frac{1}{2at}[4q_2 - q_1]$$

or

$$\frac{2}{\pi}[(\pi + 2)q_1 - 2q_2] = \frac{1}{2}[4q_2 - q_1]$$

$$\frac{4}{\pi}[(\pi + 2)q_1 - 2q_2] = [4q_2 - q_1]$$

$$\therefore \frac{4(\pi + 2)}{\pi}q_1 - \frac{8}{\pi}q_2 - 4q_2 + q_1 = 0$$

$$\left[\frac{4(\pi + 2)}{\pi} + 1\right]q_1 - \left[\frac{8}{\pi} + 4\right]q_2 = 0$$

$$\left[\frac{4(\pi + 2) + \pi}{\pi}\right]q_1 - \left[\frac{8 + 4\pi}{\pi}\right]q_2 = 0$$

or

$$(4\pi + 8 + \pi)q_1 = (8 + 4\pi)q_2$$

$$\therefore q_2 = \left(\frac{5\pi + 8}{4\pi + 8}\right)q_1$$

But the torque due to shear flows should be equal to the applied torque.

i.e.

$$T = 2q_1 A_1 + 2q_2 A_2 \qquad\qquad\text{(c)}$$

Substituting the values of q_2, A_1 and A_2 in (c), we get,

$$T = 2q_1\left(\frac{\pi a^2}{2}\right) + 2\left(\frac{5\pi + 8}{4\pi + 8}\right)q_1.4a^2$$

(a)   **(b)**

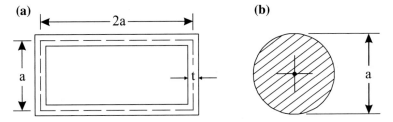

Fig. 7.14 Thin-walled box section

$$= \pi a^2 q_1 + 8a^2 \left(\frac{5\pi + 8}{4\pi + 8} \right) q_1$$

$$\therefore T = \left[\frac{a^2 \left(\pi^2 + 12\pi + 16 \right)}{(\pi + 2)} \right] q_1$$

$$\therefore q_1 = \frac{(\pi + 2)T}{a^2 \left(\pi^2 + 12\pi + 16 \right)}$$

Now, from Eq. (7.1), we have,

$$2G\theta = \frac{2}{\pi a t} \left[(\pi + 2) \frac{(\pi + 2)T}{a^2 \left(\pi^2 + 12\pi + 16 \right)} - 2 \left(\frac{5\pi + 8}{4\pi + 8} \right) \frac{(\pi + 2)T}{a^2 \left(\pi^2 + 12\pi + 16 \right)} \right]$$

Simplifying, we get the twist as $\theta = \left[\frac{(2\pi + 3)T}{2Ga^3 t \left(\pi^2 + 12\pi + 16 \right)} \right]$.

Example 7.4 A thin-walled box section having dimensions $2a \times a \times t$ is to be compared with a solid circular section of diameter as shown in Fig. 7.14. Determine the thickness t so that the two sections have:

(a) Same maximum shear stress for the same torque.
(b) The same stiffness.

Solution

(a) For the box section, we have

$$T = 2q A$$
$$= 2.\tau.t.A$$
$$T = 2.\tau.t.2a \times a$$
$$\therefore \tau = \frac{T}{4a^2 t} \qquad\qquad\qquad (i)$$

Now, for solid circular section, we have

$$\frac{T}{I_p} = \frac{\tau}{r}$$

where I_p = polar moment of inertia

$$\therefore \frac{T}{\left(\frac{\pi a^4}{32}\right)} = \frac{\tau}{\left(\frac{a}{2}\right)}$$

or

$$\frac{32T}{\pi a^4} = \frac{2\tau}{a}$$

$$\therefore \tau = \left(\frac{16T}{\pi a^3}\right) \qquad \text{(ii)}$$

Equating (i) and (ii), we get

$$\frac{T}{4a^2 t} = \frac{16T}{\pi a^3} \qquad \therefore 64a^2 t T = \pi a^3 T$$

$$\therefore t = \frac{\pi a}{64}$$

(b)   The stiffness of the box section is given by

$$\theta = \frac{q}{2GA} \oint \frac{ds}{t}$$

Here

$$T = 2qA \therefore q = \frac{T}{2A}$$

$$\therefore \theta = \frac{T}{4GA^2}\left[\frac{a}{t} + \frac{2a}{t} + \frac{a}{t} + \frac{2a}{t}\right] as$$

$$= \frac{6aT}{4GA^2 t}$$

$$= \frac{6aT}{4G\left(2a^2\right)^2 t}$$

$$\therefore \theta = \frac{6aT}{16a^4 Gt} \qquad \text{(iii)}$$

The stiffness of the solid circular section is

$$\theta = \frac{T}{GI_p} = \frac{T}{G\left(\frac{\pi a^4}{32}\right)} = \frac{32T}{G\pi a^4} \qquad \text{(iv)}$$

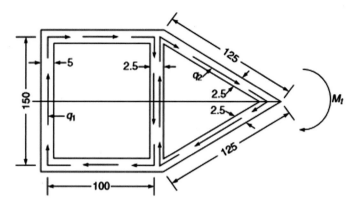

Fig. 7.15 Two-cell tube

Equating (iii) and (iv), we get

$$\frac{6aT}{16a^4Gt} = \frac{32T}{G\pi a^4}$$

$$\frac{6a}{16t} = \frac{32}{\pi}$$

$$\therefore t = \frac{6\pi a}{16 \times 32}$$

$$\therefore t = \frac{3}{4}\left(\frac{\pi a}{64}\right)$$

Example 7.5 A two-cell tube as shown in Fig. 7.15 is subjected to a torque of 10 kNm. Determine the shear stress in each part and angle of twist per metre length. Take modulus of rigidity of the material as 83 kN/mm².

Solution For Cell 1.

$$\text{Area of the Cell} = A_1 = 150 \times 100 = 15{,}000 \, \text{mm}^2$$

$$a_1 = \oint \frac{ds}{t} \text{(including web)}$$

$$= \frac{150}{5} + \frac{100}{5} + \frac{150}{2.5} + \frac{100}{5}$$

$$= 130$$

For Cell 2

$$\text{Area of the cell} = A_2 = \frac{1}{2} \times 150 \times \sqrt{(125)^2 - (75)^2}$$

$$= 7500 \, \text{mm}^2$$

$$\therefore a_2 = \oint \frac{ds}{t} \text{(including web)}$$

$$= \frac{150}{2.5} + \frac{125}{2.5} + \frac{125}{2.5}$$

$$\therefore a_2 = 160$$

For the web,

$$a_{12} = \frac{150}{2.5} = 60$$

For Cell (1)

$$2G\theta = \frac{1}{A_1}(a_1 q_1 - a_{12} q_2)$$

$$\therefore 2G\theta = \frac{1}{15,000}(130 q_1 - 60 q_2) \tag{a}$$

For Cell (2)

$$2G\theta = \frac{1}{A_2}(a_2 q_2 - a_{12} q_1)$$

$$= \frac{1}{7500}(160 q_2 - 60 q_1) \tag{b}$$

Equating (a) and (b), we get

$$\frac{1}{15,000}(130 q_1 - 60 q_2) = \frac{1}{7500}(160 q_2 - 60 q_1)$$

Solving,

$$q_1 = 1.52 q_2 \tag{c}$$

Now, the torque due to shear flows should be equal to the applied torque.

i.e.,

$$M_t = 2q_1 A_1 + 2q_2 A_2$$

$$10 \times 10^6 = 2q_1(15,000) + 2q_2(7500) \tag{d}$$

Substituting (c) in (d), we get

$$10 \times 10^6 = 2 \times 15,000(1.52 q_2) + 2q_2(7500)$$

$$\therefore q_2 = 165.02N$$

$$\therefore q_1 = 1.52 \times 165.02 = 250.83N$$

Shear flow in the web $= q_3 = (q_1 - q_2) = (250.83 - 165.02)$

$$\therefore q_3 = 85.81N$$

$$\therefore \tau_1 = \frac{q_1}{t_1} = \frac{250.83}{5} = 50.17\,\text{N/mm}^2$$

$$\tau_2 = \frac{q_2}{t_2} = \frac{165.02}{2.5} = 66.01\,\text{N/mm}^2$$

$$\tau_3 = \frac{q_3}{t_3} = \frac{85.81}{2.5} = 34.32\,\text{N/mm}^2$$

Now, the twist θ is computed by substituting the values of q_1 and q_2 in equation (a).

i.e.,

$$2G\theta = \frac{1}{15{,}000}[130 \times 250.83 \times 60 \times 165.02]$$

$$\therefore \theta = \frac{1}{15{,}000} \times \frac{22{,}706.7}{83 \times 1000} = 1.824 \times 10^{-5}\,\text{rad/mm length}$$

or

$$\theta = 1.04°/\text{m length}$$

Example 7.6 A tubular section having three cells as shown in Fig. 7.16 is subjected to a torque of 113 kNm. Determine the shear stresses developed in the walls of the section.

Solution Let $q_1, q_2, q_3, q_4, q_5, q_6$ be the shear flows in the various walls of the tube as shown in the figure. A_1, A_2, and A_3 be the areas of the three cells.

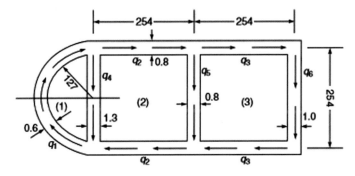

Fig. 7.16 Three cell with tubular section

$$\therefore A_1 = \frac{\pi}{2}(127)^2 = 25,322\,\text{mm}^2$$

$$A_2 = 254 \times 254 = 64,516\,\text{mm}^2$$

$$A_3 = 64,516\,\text{mm}^2$$

Now, From the figure,

$$q_1 = q_2 + q_4$$

$$q_2 = q_3 + q_5$$

$$q_3 = q_5$$

or

$$q_1 = \tau_1 t_1 = \tau_2 t_2 + \tau_4 t_4$$

$$q_2 = \tau_2 t_2 = \tau_3 t_3 + \tau_5 t_5$$

$$q_3 = \tau_3 t_3 = \tau_6 t_6 \tag{1}$$

where $\tau_1, \tau_2, \tau_3, \tau_4, \tau_5$ and τ_6 are the shear stresses in the various walls of the tube.
Now, the applied torque is

$$M_t = 2A_1 q_1 + 2A_2 q_2 + 2A_3 q_3$$

$$= 2(A_1 \tau_1 t_1 + A_2 \tau_2 t_2 + A_3 \tau_3 t_3)$$

i.e.,

$$113 \times 10^6 = 2[(25,322\tau_1 \times 0.8) + (64,516\tau_2 \times 0.8) + (64,516 \times 0.8)]$$

$$\therefore \tau_1 + 3.397(\tau_2 + \tau_3) = 3718 \tag{2}$$

Now, considering the rotations of the cells and S_1, S_2, S_3, S_4, S_5 and S_6 as the length of cell walls,
We have,

$$\tau_1 S_1 + \tau_4 S_4 = 2G\theta A_1$$

$$-\tau_4 S_4 + 2\tau_2 S_2 + \tau_5 S_5 = 2G\theta A_2$$

$$-\tau_5 S_5 + 2\tau_3 S_3 + \tau_6 S_6 = 2G\theta A_3 \tag{3}$$

Here

$$S_1 = (\pi \times 127) = 398\,\text{mm}$$

$$S_2 = S_3 = S_4 = S_5 = S_6 = 254\,\text{mm}$$

∴(3) can be written as

$$398\tau_1 + 254S_4 = 25{,}322G\theta$$
$$-254\tau_2 + 2 \times 254 \times \tau_2 + 254\tau_5 = 64{,}516G\theta$$
$$-254\tau_2 + 2 \times 254 \times \tau_3 + 254\tau_6 = 64{,}516G\theta \tag{4}$$

Now, solving (1), (2) and (4) we get

$$\tau_1 = 40.4\,\text{N/mm}^2$$
$$\tau_2 = 55.2\,\text{N/mm}^2$$
$$\tau_3 = 48.9\,\text{N/mm}^2$$
$$\tau_4 = -12.7\,\text{N/mm}^2$$
$$\tau_6 = 36.6\,\text{N/mm}^2$$

7.11 Exercises

1. Develop expression for the shear stress and torsional rigidity of single cell tube under torque.
2. The cross-section of hollow prismatic tube is as shown in Fig. 7.17. Evaluate the angle of twist per unit length produced taking rigidity modulus as 75 GPa. Also evaluate the maximum shear stress produced.

3. A hollow thin-wall torsion member has two compartments as shown in Fig. 7.18. The material is aluminium alloy for which $G = 26$ GPa. Determine the torque and unit angle of the twist if the maximum shearing stress is 40 MPa.

4. The hollow circular and square thin-wall members are having identical values of b and t as shown in Fig. 7.19. Determine the ratio of the torques and unit angle of twists for the two torsion members if shearing stresses are equal in both sections. Neglect the effect of stress concentration.

5. A hollow thin-wall member has dimensions as shown in Fig. 7.20. It has a total

Fig. 7.17 Hollow prismatic tube

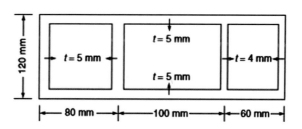

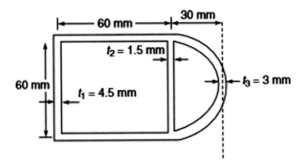

Fig. 7.18 Hollow thin-wall torsion member

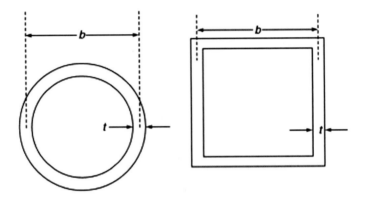

Fig. 7.19 Hollow circular and square thin-wall members

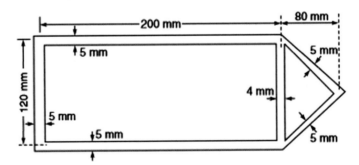

Fig. 7.20 Hollow thin-wall member

length of 3000 mm. The member is subjected to a torque of 1100 kNm. Determine the maximum shearing stress and the angle of twist.

6. A tubular aluminium shaft having a square cross-section with outside dimension 25 mm must safely carry a twisting moment of 69.5 Nm. Calculate proper wall thickness t if the working stress in shear is 42 MN/m^2.

7. Find the torsional moment capacity of a thin-walled square box Section 200 mm × 200 mm with 5 mm walls if the shear stress is 80 N/mm^2. Find also the mean diameter of a circular tube with the same thickness to resist the same twisting moment and shear stress.

8. A thin-walled box section having dimensions shown in figure is subjected to a torque of 15 kNm. Calculate the shear stresses developed in each part and the angle of twist per metre length if $G = 84$ kN/m^2 (Fig. 7.21).

9. A box section is shown in Fig. 7.22. Prove that for a symmetrical section such as this, there is no stress in the central web when the tube is twisted. Derive an expression for the shear stress in the section.

10. An aeroplane elevator has dimensions as shown in Fig. 7.23. Calculate the distribution of shear stresses in the elevator. If the elevator is made of an aluminium alloy, $G = 26.7$ kN/mm^2 and is 1524 mm long, what is the total angle of twist.

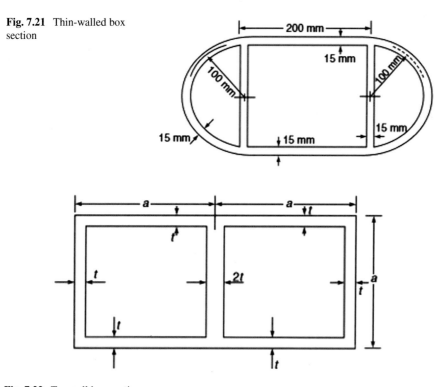

Fig. 7.21 Thin-walled box section

Fig. 7.22 Two cell box section

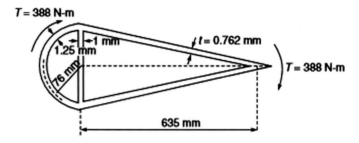

Fig. 7.23 Aeroplane elevator

Fig. 7.24 Three cell tube
section

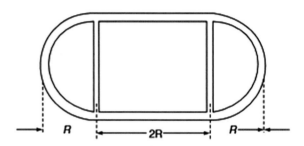

11. Show that for the same twist, the elliptic section has a great shearing stress than the inscribed circular section (radius equal to the minor axis b of the ellipse) that takes the greater torque for the same allowable stress.

12. Find the expressions for the shear stresses in a tube of the section shown in Fig. 7.24, taking wall thickness as t uniform throughout.

Chapter 8
Elastic Solutions in Geomechanics

8.1 Introduction

Many of the elastic solutions for problems in geomechanics were solved in the
later part of nineteenth century, and they were usually solved not for geotechnical
engineering applications, but to answer basic questions about elasticity and behaviour
of elastic bodies. Some special problems that hold a fundamental position in relation
to elastic solutions in geotechnical engineering are presented in this chapter. With
one exception, all the problems presented involve a point load. This point load is a
finite force applied at a point (on a surface of zero area). Due to stress singularities,
the point load problems involve limiting procedures. Solutions of the problems of
Kelvin, Boussinesq, Flamant Cerrutti and Mindlin related to point load are discussed
in this chapter.

The classical problem of Kelvin involves a point load acting in the interior of
infinite elastic body. The Kelvin's solution has found many applications in engineering. This is useful for the analysis of screw plate in which the plate is embedded
in an infinite elastic body and the screw plate can be represented by an assumed
load distribution. The classical problem of Boussinesq dealing with a normal force
applied at the plane boundary of a semi-infinite solid has found practical application
in the study of the distribution of foundation pressures, contact stresses and in other
problems of soil mechanics.

Of all the above problems discussed in this chapter, the most useful in geomechanics is the Boussinesq's solution. The Flamant's problem uses Boussinesq's solution together with the principle of superposition to solve for a stress field for a line
load acting on the surface of the half-space. The Flamant problem has many applications in determining the stress and displacement fields below continuous strip
footings, embankments and retaining wall foundations. The fundamental solution of
Cerrutti for horizontal load acting at the surface of the elastic half-space has found
applications for stress and displacement fields in geotechnical engineering dealing
with lateral loads at the ground surface. Examples include single or group of piles
subjected to horizontal lateral load at the ground line. The fundamental solution by

© The Author(s), under exclusive license to Springer Nature Singapore Pte Ltd. 2021   261
T. G. Sitharam and L. Govindaraju, *Theory of Elasticity*,
https://doi.org/10.1007/978-981-33-4650-5_8

Mindlin involved with point load (either vertical or horizontal) acting in the interior of an elastic half-space. This solution has found many applications to determine the stresses and displacement fields surrounding an axially loaded pile and in the study of interactions between foundations and ground anchors. The following sections describe the selected fundamental solutions in geomechanics.

8.2   Kelvin's Problem

The problem is to determine the stresses in an infinite elastic body when a point load is acting in the interior of a body (as shown in Fig. 8.1).

A point load of magnitude 2P is considered to be acting at a point in the interior of an infinite elastic body. In the cylindrical co-ordinate system, the following displacements can be obtained by Kelvin's solution (Kelvin 1848).

Displacement in radial direction $= u_r = \frac{P_{rz}}{8\pi G(1-v)R^3}$

Tangential displacement $= u_\theta = 0$

$$\text{Vertical displacement} = u_z = \frac{P}{8\pi G(1-v)}\left[\frac{2(1-2v)}{R} + \frac{1}{R} + \frac{z^2}{R^3}\right] \quad (8.1)$$

Similarly, the stresses are given by

$$\sigma_r = -\frac{P}{4\pi(1-v)}\left[\frac{(1-2v)z}{R^3} - \frac{3r^2z}{R^5}\right]$$

$$\sigma_\theta = \frac{P(1-2v)z}{4\pi(1-v)R^3}$$

$$\sigma_z = \frac{P}{4\pi(1-v)}\left[\frac{(1-2v)z}{R^3} + \frac{3z^3}{R^5}\right] \quad (8.2)$$

$$\tau_{rz} = \frac{P}{4\pi(1-v)}\left[\frac{(1-2v)r}{R^3} + \frac{3rz^2}{R^5}\right]$$

Fig. 8.1 Kelvin's Problem

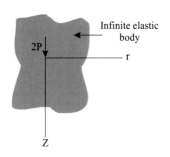

Infinite elastic body

2P

r

Z

$$\tau_{r\theta} = \tau_{\theta r} = \tau_{\theta z} = \tau_{z\theta} = 0$$

Here, $R = \sqrt{z^2 + r^2}$

From the above expressions, it is clear that both displacements and stresses reduce to zero for larger values of R. However, on the plane $z = 0$, all the stress components except for τ_{rz} vanish, at all points except the origin.

Vertical Tractions Equilibrating the Applied Point Load

A planar surface at $z = h$ is considered (as shown in Fig. 8.2), and the the vertical component of traction on this surface is σ_z. integrating σ_z; over this entire surface, one can get the resultant force. To find this resultant force, consider a horizontal circle centred on the z-axis over which σ_z is constant (Fig. 8.3). Hence, the force acting on the annulus shown in Fig. 8.3 will be $\sigma_z \times 2\pi r dr$.

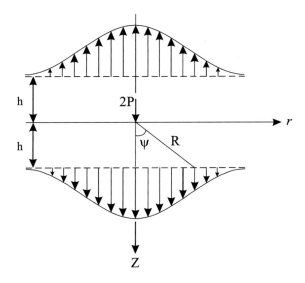

Fig. 8.2 Vertical stress distribution on horizontal planes above and below point load

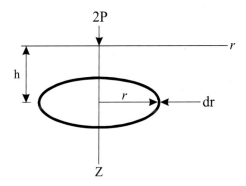

Fig. 8.3 Geometry for integrating vertical stress

Now, the total resultant force on the surface $z = h$ is given by,

$$\text{Resultant upward force} = \int_0^\infty \sigma_z(2\pi r dr)$$

$$= \int_0^\infty \frac{P}{4\pi(1-v)}\left[\frac{(1-2v)h}{R^3} + \frac{3h^3}{R^5}\right](2\pi r dr)$$

$$= \int_0^\infty \frac{P}{2(1-v)}\left[\frac{(1-2v)h}{R^3} + \frac{3h^3}{R^5}\right]r\, dr$$

To simplify the integration, introduce the angle ψ as shown in Fig. 8.2.
Here, $r = h\tan\psi$ and $dr = h\sec^2\psi\, d\psi$

$$\text{Therefore, resultant upward force} = \int_0^{\pi/2} \frac{P}{2(1-v)}\left[(1-2v)\sin\psi + 3\cos^2\psi\sin\psi\right]d\psi$$

$$(8.3)$$

Solving, we get resultant upward force on the lower plane $= P$ which is exactly one-half the applied load.

Further, if we consider a similar surface $z = -h$, shown in Fig. 8.3, we will find tensile stresses of the same magnitude as the compressive stresses on the lower plane.

Hence, resultant force on the upper plane $= -P$ (tensile force). Combining the two resultant forces, we get $2P$ which exactly equilibrate the applied load.

8.3 Boussinesq's Problem

The problem of a point load acting normal to the surface of an elastic half-space was solved by the French mathematician Joseph Boussinesq (1878). The problem geometry is illustrated in the following Fig. 8.4. The half-space is assumed to be homogeneous, isotropic and elastic. The point load is applied at the origin of co-ordinates on the half-space surface. Let P be the magnitude of the point load. Now, consider the stress function

$$\phi = B\left(r^2 + z^2\right)^{\frac{1}{2}} \tag{8.4}$$

where B is a constant.
    The stress components are given by

Fig. 8.4 Boussinesq's
problem

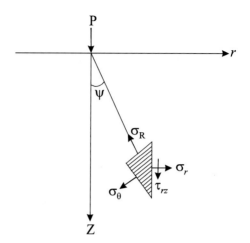

$$\sigma_r = \frac{\partial}{\partial z}\left(v\nabla^2\phi - \frac{\partial^2\phi}{\partial r^2}\right)$$

$$\sigma_\theta = \frac{\partial}{\partial z}\left(v\nabla^2\phi - \frac{1}{r}\frac{\partial\phi}{\partial r}\right) \tag{8.5}$$

$$\sigma_z = \frac{\partial}{\partial z}\left[(2-v)\nabla^2\phi - \frac{\partial^2\phi}{\partial z^2}\right]$$

$$\tau_{rz} = \frac{\partial}{\partial r}\left[(1-v)\nabla^2\phi - \frac{\partial^2\phi}{\partial z^2}\right]$$

Therefore, by substitution, we get

$$\sigma_r = B\left[(1-2v)z(r^2+z^2)^{-\frac{3}{2}} - 3r^2z(r^2+z^2)^{-\frac{5}{2}}\right]$$

$$\sigma_\theta = B[1-2v]z(r^2+z^2)^{-\frac{3}{2}} \tag{8.6}$$

$$\sigma_z = B\left[(1-2v)z(r^2+z^2)^{-\frac{3}{2}} + 3z^3(r^2+z^2)^{-\frac{5}{2}}\right]$$

$$\tau_{rz} = -B\left[(1-2v)r(r^2+z^2)^{-\frac{3}{2}} + 3rz^2(r^2+z^2))^{-\frac{5}{2}}\right]$$

Now, the shearing forces on the boundary plane $z=0$ are given by

$$\tau_{rz} = \frac{-B(1-2v)}{r^2} \tag{a}$$

In polar co-ordinates, the distribution of stress is given by

$$\sigma_R = \frac{A}{R^3}, \sigma_\theta = \sigma_R + \frac{d\sigma_R}{dR}\frac{R}{2}$$

or $\sigma_\theta = -\frac{1}{2}\frac{A}{R^3}$

where A is a constant and $R = \sqrt{r^2 + z^2}$

In cylindrical co-ordinates, we have the following expressions for the stress components:

$$\sigma_r = \sigma_R \sin^2 \psi + \sigma_\theta \cos^2 \psi$$

$$\sigma_z = \sigma_R \cos^2 \psi + \sigma_\theta \sin^2 \psi \qquad (8.7)$$

$$\tau_{rz} = \frac{1}{2}(\sigma_R - \sigma_\theta)\sin 2\psi$$

$$\sigma_\theta = -\frac{1}{2}\frac{A}{R^3}$$

But from Fig. 8.4

$$\sin \psi = r\left(r^2 + z^2\right)^{-\frac{1}{2}}$$

$$\cos \psi = z\left(r^2 + z^2\right)^{-\frac{1}{2}}$$

Substituting the above, into σ_r, σ_z, $_{rz}$ and σ_θ, we get

$$\sigma_r = A\left(r^2 - \frac{1}{2}z^2\right)\left(r^2 + z^2\right)^{-\frac{5}{2}}$$

$$\sigma_z = A\left(z^2 - \frac{1}{2}r^2\right)\left(r^2 + z^2\right)^{-\frac{5}{2}}$$

$$\tau_{rz} = \frac{3}{2}\left(r^2 + z^2\right)^{-\frac{5}{2}}(A\,r\,z) \qquad (8.8)$$

$$\sigma_\theta = -\frac{1}{2}A\left(r^2 + z^2\right)^{-\frac{3}{2}}$$

Considering that centres of pressure are uniformly distributed along the z-axis from $z = 0$ to $z = -\infty$, then by superposition, the stress components produced are given by

$$\sigma_r = A_1 \int_z^\infty \left(r^2 - \frac{1}{2}z^2\right)\left(r^2 + z^2\right)^{-\frac{5}{2}} dz$$

$$= \frac{A_1}{2}\left[\frac{1}{r^2} - \frac{z}{r^2}\left(r^2 + z^2\right)^{-\frac{1}{2}} - z\left(r^2 + z^2\right)^{-\frac{3}{2}}\right]$$

$$\sigma_z = A_1 \int_z^\infty \left(z^2 - \frac{1}{2}r^2\right)\left(r^2 + z^2\right)^{-\frac{5}{2}} dz$$

$$= \frac{A_1}{2}z\left(r^2 + z^2\right)^{-\frac{3}{2}} \tag{8.9}$$

$$\tau_{rz} = \frac{3}{2}A_1 \int_z^\infty rz\left(r^2 + z^2\right)^{-\frac{5}{2}} dz = \frac{A_1}{2}r\left(r^2 + z^2\right)^{-\frac{3}{2}}$$

$$\sigma_\theta = -\frac{1}{2}A_1 \int_z^\infty \left(r^2 + z^2\right)^{-\frac{3}{2}} dz$$

$$= -\frac{A_1}{2}\left[\frac{1}{r^2} - \frac{z}{r^2}\left(r^2 + z^2\right)^{-\frac{1}{2}}\right]$$

On the plane $z = 0$, we find that the normal stress is zero and the shearing stress is

$$\tau_{rz} = \frac{1}{2}\frac{A_1}{r^2} \tag{b}$$

From (a) and (b), it is seen that the shearing forces on the boundary plane are eliminated if, $-B(1 - 2v) + \frac{A_1}{2} = 0$

Therefore, $A_1 = 2B(1 - 2v)$

Substituting the value of A_1 in Eq. (8.9) and adding together the stresses from Eqs. (8.6) and (8.9), we get

$$\sigma_r = B\left\{(1 - 2v)\left[\frac{1}{r^2} - \frac{z}{r^2}\left(r^2 + z^2\right)^{-\frac{1}{2}}\right] - 3r^2z\left(r^2 + z^2\right)^{-\frac{5}{2}}\right\}$$

$$\sigma_z = -3Bz^3\left(r^2 + z^2\right)^{-\frac{5}{2}}$$

$$\sigma_\theta = B(1 - 2v)\left[-\frac{1}{r^2} + \frac{z}{r^2}\left(r^2 + z^2\right)^{-\frac{1}{2}} + z\left(r^2 + z^2\right)^{-\frac{3}{2}}\right]$$

$$\tau_{rz} = -3Brz^2\left(r^2 + z^2\right)^{-\frac{5}{2}} \tag{8.10}$$

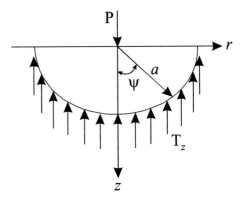

Fig. 8.5 Vertical tractions acting on the hemispherical surface

The above stress distribution satisfies the boundary conditions, since $\sigma_z = \tau_{rz} = 0$ for $z = 0$.

To Determine the Constant B

Consider the hemispherical surface of radius "a" as illustrated in Fig. 8.5. For any point on this surface, let $R = a = $ constant. Also, ψ be the angle between a radius of the hemisphere and the z-axis.

The unit normal vector to the surface at any point can be written as

$$\hat{n} = \begin{bmatrix} \sin \psi \\ 0 \\ \cos \psi \end{bmatrix}$$

while r and z components of the point are

$$z = a \cos \psi, r = a \sin \psi$$

The traction vector that acts on the hemispherical surface is,

$$T = \begin{bmatrix} T_r \\ T_\theta \\ T_z \end{bmatrix} = \begin{bmatrix} \sigma_r & 0 & \tau_{rz} \\ 0 & \sigma_\theta & 0 \\ \tau_{rz} & 0 & \sigma_z \end{bmatrix} \begin{bmatrix} \sin \psi \\ 0 \\ \cos \psi \end{bmatrix} = \begin{bmatrix} \sigma_r \sin \psi + \tau_{rz} \cos \psi \\ 0 \\ \tau_{rz} \sin \psi + \sigma_z \cos \psi \end{bmatrix}$$

Considering the component of stress in the z-direction on the hemispherical surface,

we can write

$$T_z = -(\tau_{rz} \sin \psi + \sigma_z \cos \psi)$$

Substituting the values of τ_{rz}, z, $\sin\psi$ and $\cos\psi$, in the above expression, we get

$$T_z = 3Bz^2(r^2 + z^2)^{-2}$$

Integrating the above, we get the applied load P.
Therefore,

$$P = \int_z^{\pi/2} T_z 2\pi r(r^2 + z^2)^{\frac{1}{2}} d\psi$$

$$= \int_z^{\pi/2} \left[3Bz^2(r^2 + z^2)^{-2}\right]\left[2\pi r(r^2 + z^2)^{\frac{1}{2}}\right] d\psi$$

$$= \int_z^{\pi/2} (6\pi B)\cos^2\psi \sin\psi d\psi$$

$$= 6\pi B \int_0^{\pi/2} \cos^2\psi \sin\psi d\psi$$

Now, solving for $\int_0^{\pi/2} \cos^2\psi \sin\psi d\psi$, we proceed as below

$$\text{Put } \cos\psi = t$$

i.e. $-\sin\psi$, $d\psi = dt$
   If $\cos 0 = 1$, then $t = 1$
   If $\cos\frac{\pi}{2} = 0$, then $t = 0$
   Hence, $\int_0^1 -t^2 dt = -\left[\frac{t^3}{3}\right]_1^0 = \left[\frac{t^3}{3}\right]_0^1 = \frac{1}{3}$ Therefore, $P = 2\pi B$
   Or $B = \frac{P}{2\pi}$
   Substituting the value of B in Eq. (8.10), we get

$$\sigma_z = -\frac{3P}{2\pi} z^3(r^2 + z^2)^{-\frac{5}{2}}$$

$$\sigma_r = \frac{P}{2\pi}\left\{(1 - 2v)\left[\frac{1}{r^2} - \frac{z}{r^2}(r^2 + z^2)^{-\frac{1}{2}}\right] - 3r^2 z(r^2 + z^2)^{-\frac{5}{2}}\right\}$$

$$\sigma_\theta = \frac{P}{2\pi}(1 - 2v)\left\{-\frac{1}{r^2} + \frac{z}{r^2}(r^2 + z^2)^{-\frac{1}{2}} + z(r^2 + z^2)^{-\frac{3}{2}}\right\}$$

$$\tau_{rz} = -\frac{3P}{2\pi} rz^2(r^2 + z^2)^{-\frac{5}{2}}$$

Putting $R = \sqrt{r^2 + z^2}$ and simplifying, we can write

$$\sigma_z = -\frac{3Pz^3}{2\pi R^5}$$

$$\sigma_r = \frac{P}{2\pi}\left[\frac{(1-2v)}{R(R+z)} - \frac{3r^2z}{R^5}\right]$$

$$\sigma_\theta = \frac{P(1-2v)}{2\pi}\left[\frac{z}{R^3} - \frac{1}{R(R+z)}\right]$$

$$\tau_{rz} = -\frac{3P}{2\pi}\frac{rz^2}{R^5}$$

For the loading considered, Boussinesq gave the following solution for displacements $u_r = \frac{P}{4\pi GR}\left[\frac{rz}{R^2} - \frac{(1-2v)r}{R+z}\right]$

$$u_\theta = 0$$

$$u_z = \frac{P}{4\pi GR}\left[2(1-v) + \frac{z^2}{R^2}\right]$$

Comparison Between Kelvin's and Boussinesq's solutions

On the plane $z = 0$, all the stresses given by Kelvin vanish except τ_{rz}. For the special case where Poisson's ratio $v = 0.5$ (an incompressible material), then τ_{rz} will also be zero on this surface, and that part of the body below the $z = 0$ plane becomes equivalent to the half-space of Boussinesq's problem. Comparing Kelvin's solution (with $v = 0.5$) with Boussinesq's solution (with $v = 0.5$), it is clear that for all $z \geq 0$, the solutions are identical. For $z \leq 0$, we also have Boussinesq's solution, but with a negative load $-P$. The two half-spaces, which together comprise the infinite body of Kelvin's problem, act as if they are uncoupled on the plane $z = 0$, where they meet.

Further, a spherical surface is centred on the origin, we find a principal surface on which the major principal stress is acting. The magnitude of the principal stress is given by

$$\sigma_1 = \frac{3Pz}{2\pi R^3} \tag{8.11}$$

where R is the sphere radius. It can be observed that the value of σ_1 changes for negative values of z, giving tensile stresses above the median plane $z = 0$.

8.4 Flamant's Problem

Consider a case of a line load of intensity "q" per unit length acting on the surface of a homogeneous, elastic and isotropic half-space as shown in Fig. 8.6.

The stresses at a point $P(r,)$ can be determined by using the stress function

$$\phi = \frac{q}{\pi} r\theta \sin\theta \qquad (8.12)$$

In the polar co-ordinate system, the expressions for the stresses are as follows:

$$\sigma_r = \frac{1}{r}\frac{\partial\phi}{\partial r} + \frac{1}{r^2}\frac{\partial^2\phi}{\partial\theta^2} \qquad (8.13)$$

and

$$\sigma_\theta = \frac{\partial^2\phi}{\partial r^2} \qquad (8.14)$$

$$\tau_{r\theta} = -\frac{\partial}{\partial r}\left(\frac{1}{r}\frac{\partial\phi}{\partial\theta}\right) \qquad (8.15)$$

Now, differentiating Eq. (8.12) with respect to r, we get

$$\frac{\partial\phi}{\partial r} = \frac{q}{\pi}\theta\sin\theta$$

Similarly $\frac{\partial^2\phi}{\partial r^2} = 0$ and $\sigma_\theta = 0$

Also, differentiating Eq. (8.12) with respect to θ, we get

Fig. 8.6 Vertical line load on surface of an half-space

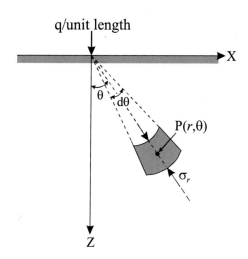

$$\frac{\partial \phi}{\partial \theta} = \frac{q}{\pi} r [\theta \cos \theta + \sin \theta]$$

$$\frac{\partial \phi}{\partial \theta} = \frac{qr\theta \cos \theta}{\pi} + \frac{qr \sin \theta}{\pi}$$

Similarly, $\frac{\partial^2 \phi}{\partial \theta^2} = \frac{qr}{\pi}[\theta(-\sin \theta) + \cos \theta] + \frac{qr}{\pi} \cos \theta$

$$\frac{\partial^2 \phi}{\partial \theta^2} = \frac{qr}{\pi} \cos \theta + \frac{qr}{\pi} \cos \theta - \frac{qr\theta}{\pi} \sin \theta$$

Therefore, Eq. (8.13) becomes

$$\sigma_r = \frac{1}{r}\left(\frac{q}{\pi}\theta \sin \theta\right) + \frac{1}{r^2}\left(\frac{q}{\pi}r \cos \theta + \frac{q}{\pi}r \cos \theta - \frac{q}{\pi}r\theta \sin \theta\right)$$

$$= \frac{q}{\pi r}\theta \sin \theta + \frac{2q}{\pi r}\cos \theta - \frac{q}{\pi r}\theta \sin \theta$$

Or $\sigma_r = \frac{2q \cos \theta}{\pi r}$

Now, $\frac{1}{r}\frac{\partial \phi}{\partial \theta} = \frac{q}{\pi}[\theta \cos \theta + \sin \theta]$

$$\frac{\partial}{\partial r}\left(\frac{1}{r}\frac{\partial \phi}{\partial \theta}\right) = 0$$

Hence, $\tau_{r\theta} = 0$

The stress function assumed in Eq. (8.12) will satisfy the compatibility equation

$$\left(\frac{\partial^2}{\partial r^2} + \frac{1}{r}\frac{\partial}{\partial r} + \frac{1}{r^2}\frac{\partial^2}{\partial \theta^2}\right)\left(\frac{\partial^2 \phi}{\partial r^2} + \frac{1}{r}\frac{\partial \phi}{\partial r} + \frac{1}{r^2}\frac{\partial^2 \phi}{\partial \theta^2}\right) = 0$$

Here, σ_r and σ_θ are the major and minor principal stresses at point P. Now, using the above expressions for σ_r, σ_θ and $\tau_{r\theta}$, the stresses in rectangular co-ordinate system (Fig. 8.7) can be derived.

Therefore,

$$\sigma_z = \sigma_r \cos^2 \theta + \sigma_\theta \sin^2 \theta - 2\tau_{r\theta} \sin \theta \cos \theta$$

Here, $\sigma_\theta = 0$ and $\tau_{r\theta} = 0$

Hence, $\sigma_z = \sigma_r \cos^2 \theta$

$$\frac{2q}{\pi r} \cos \theta \left(\cos^2 \theta\right)$$

$$\sigma_z = \frac{2q}{\pi r} \cos^3 \theta$$

Fig. 8.7 Stresses due to a
vertical line load in
rectangular co-ordinates

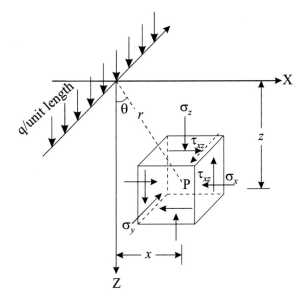

But from Fig. 8.7,

$$r = \sqrt{x^2 + z^2}$$

$$\cos \theta = \frac{z}{\sqrt{\left(x^2 + z^2\right)}}, \ \sin \theta = \frac{x}{\sqrt{x^2 + z^2}}$$

Therefore,

$$\sigma_z = \frac{2q}{\pi \sqrt{\left(x^2 + z^2\right)}} \frac{z^3}{\left(\sqrt{x^2 + z^2}\right)^3}$$

$$\sigma_{\bar{z}} = \frac{2q z^3}{\pi \left(x^2 + z^2\right)^2}$$

Similarly,

$$\sigma_x = \sigma_r \sin^2 \theta + \sigma_\theta \cos^2 \theta + 2\tau_{r\theta} \sin \theta$$

$$= \frac{2q}{\pi r} \cos \theta \sin^2 \theta + 0 + 0$$

$$= \frac{2q}{\pi \sqrt{x^2 + z^2}} \frac{z}{\sqrt{x^2 + z^2}} \frac{x^2}{\left(x^2 + z^2\right)}$$

or

$$\sigma_x = \frac{2qx^2z}{\pi \left(x^2 + z^2\right)^2}$$

and

$$\tau_{xz} = -\sigma_\theta \sin\theta \cos\theta + \sigma_r \sin\theta \cos\theta + \tau_{r\theta}\left(\cos^2\theta \sin^2\theta\right)$$

$$= 0 + \frac{2q}{\pi r} \cos\theta \sin\theta \cos\theta + 0$$

$$= \frac{2q}{\pi r} \sin\theta \cos^2\theta$$

$$= \frac{2q}{\pi r} \frac{x}{\sqrt{x^2 + z^2}} \frac{z^2}{\left(x^2 + z^2\right)}$$

or

$$\tau_{xz} = \frac{2qxz^2}{\pi \left(x^2 + z^2\right)^2}$$

But for the plane strain case,

$$\sigma_y = v(\sigma_x + \sigma_z)$$

where v = Poisson's ratio

Substituting the values of σ_x and σ_z in σ_y, we get

$$\sigma_y = \left[\frac{2qx^2z}{\pi \left(x^2 + z^2\right)^2} + \frac{2qz^3}{\pi \left(x^2 + z^2\right)^2} \right] v$$

$$= \frac{2qvz}{\pi \left(x^2 + z^2\right)} \left[x^2 + z^2\right]$$

or $\sigma_y = \frac{2qvz}{\pi \left(x^2 + z^2\right)}$

Therefore, according to Flamant's (1892) solution, the following are the stresses due to a vertical line load on the surface of an half-space.

$$\sigma_x = \frac{2qx^2z}{\pi \left(x^2 + z^2\right)^2}$$

$$\sigma_y = \frac{2qvz}{\pi \left(x^2 + z^2\right)}$$

Fig. 8.8 Cylindrical surface
aligned with line load

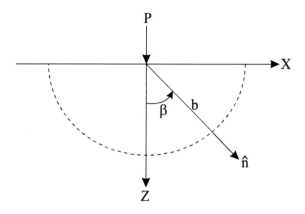

$$\sigma_z = \frac{2qz^3}{\pi \left(x^2 + z^2\right)^2} \tag{8.16}$$

$$\tau_{xz} = \frac{2qxz^2}{\pi \left(x^2 + z^2\right)^2}$$

and $\tau_{xy} = \tau_{yx} = \tau_{zy} = \tau_{yz} = 0$

Tractions acting on the Cylindrical Surface under the Line Load
One can carry out an analysis to find the tractions that act on the cylindrical surface
by using the stress components in Eq. (8.16) (Fig. 8.8).
    Here,

$$\text{the traction vector is given by } T = \frac{2qz}{\pi b^2}\hat{n} \tag{8.17}$$

where $\hat{n}$ is the unit normal to the cylindrical surface. This means to say that the
cylindrical surface itself is a principal surface. The major principal stress acts on it.
    Hence,

$$\sigma_1 = \frac{2qz}{\pi b^2} \tag{8.18}$$

The intermediate principal surface is defined by $\hat{n} = \{0, 1, 0\}$, and the intermediate
principal stress is $\sigma_2 = \nu \sigma_1$.
    The minor principal surface is perpendicular to the cylindrical surface and to
the intermediate principal surface, and the minor principal stress is exactly zero.
The other interesting characteristic of Flamant's problem is the distribution of the
principal stress in space.
    Now, consider the locus of points on which the major principal stress σ_1 is a
constant. From Eq. (8.18), this will be a surface for which

Fig. 8.9 Pressure bulb on
which the principal stresses
are constant

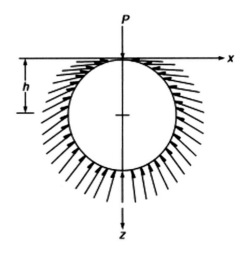

$$\frac{z}{b^2} = \frac{\pi\sigma_1}{2q} = \frac{1}{2h}$$

where h is a constant.

But $b^2 = x^2 + z^2$

Therefore, $\frac{z}{(x^2+z^2)} = \frac{1}{2h}$

or $b^2 = (x^2 + z^2) = 2\,h\,z$

which is the equation of a circle with radius h centred on the z-axis at a depth h beneath the origin, as shown in Fig. 8.9.

At every point on the circle, the major principal stress is the same. It points directly at the origin. If a larger circle is considered, the value of σ_1 would be smaller. This result gives us the idea of a "pressure bulb" in the soil beneath a foundation.

8.5   Cerrutti's Problem

This is a more complicated problem than Boussinesq's or Kelvin due to the absence of radial symmetry. Figure 8.10 shows a horizontal point load P acting on the surface of a semi-infinite soil mass.

The point load represented by P acts at the origin of co-ordinates, pointing in the x-direction. Due to the absence of symmetry, a rectangular co-ordinate system is used in the solution.

According to Cerrutti's solution (Cerrutti 1884), the displacements are given by

$$u_x = \frac{P}{4\pi GR}\left\{1 + \frac{x^2}{R^2} + (1-2v)\left[\frac{R}{R+z} - \frac{x^2}{(R+z)^2}\right]\right\} \qquad (8.19)$$

Fig. 8.10 Cerrutti's problem

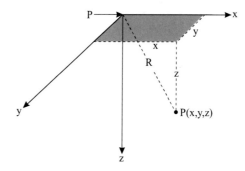

$$u_y = \frac{P}{4\pi G R}\left\{ \frac{xy}{R^2} - (1-2v)\frac{xy}{(R+z)^2} \right\} \tag{8.19a}$$

$$u_z = \frac{P}{4\pi G R}\left\{ \frac{xz}{R^2} + (1-2v)\frac{x}{R+z} \right\} \tag{8.19b}$$

and the stresses are

$$\sigma_x = -\frac{Px}{2\pi R^3}\left\{ -\frac{3x^2}{R^2} + \frac{(1-2v)}{(R+z)^2}\left[R^2 - y^2 - \frac{2Ry^2}{(R+z)} \right] \right\} \tag{8.20}$$

$$\sigma_y = -\frac{Px}{2\pi R^3}\left\{ -\frac{3y^2}{R^2} + \frac{(1-2v)}{(R+z)^2}\left[3R^2 - x^2 - \frac{2Rx^2}{(R+z)} \right] \right\} \tag{8.20a}$$

$$\sigma_z = \frac{3Pxz^2}{2\pi R^5} \tag{8.20b}$$

$$\tau_{xy} = -\frac{Py}{2\pi R^3}\left\{ -\frac{3x^2}{R^2} + \frac{(1-2v)}{(R+z)^2}\left[-R^2 + x^2 + \frac{2Rx^2}{(R+z)} \right] \right\} \tag{8.20c}$$

$$\tau_{yz} = \frac{3Pxyz}{2\pi R^5} \tag{8.20d}$$

$$\tau_{zx} = \frac{3Px^2z}{2\pi R^5} \tag{8.20e}$$

Here, $R^2 = x^2 + y^2 + z^2$

From the above, it is clear that the stresses approach to zero for large value of R. As seen from the x-component of the displacement field, it is observed that the particles are displaced in the direction of the point load. The y-component of displacement moves particles away from the x-axis for positive values of x and towards the x-axis for negative x. The plot of horizontal displacement vectors at the surface $z = 0$ is shown in Fig. 8.11 for the special case of an incompressible material. Vertical

Fig. 8.11 Distribution of
horizontal displacements
surrounding the point load

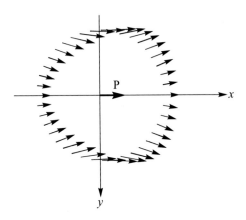

displacements take the sign of x, and hence, particles move downwards in front of
the load and upwards behind the load.

8.6   Mindlin's Problem

The two variations of the point load problem were solved by Mindlin (1936). These
are the problems of either vertical or horizontal point load acting in the interior of
an elastic
    half-space. Mindlin's problem is illustrated in Fig. 8.12. The load P acts at a point
located a distance $z = 2\,h$ beneath the half-space surface. Mindli's problem is more
complex than Boussinesq's or Kelvin's or Cerrutti's

Fig. 8.12 Mindlin's problem

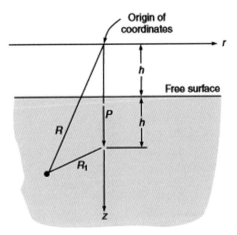

It is appropriate to write Mindlin's solution by placing the origin of co-ordinates a distance h above the free surface as shown in Fig. 8.12. Then, the applied load acts at the point $z = 2\,h$.

From Fig. 8.12,

$$R^2 = r^2 + z^2$$

$$R_1^2 = r^2 + z_1^2$$

where $z_1 = z - 2\,h\,z$

Here, z_1 and R_1 are the vertical distance and the radial distance from the point load.

For the vertical point load, Mindlin's solution is most conveniently stated in terms of Boussinesq's solution. For example, consider the displacement and stress fields in Boussinesq's problem in the region of the half-space below the surface $z = h$. These displacements and stresses are also found in Mindlin's solution, but with additional terms. The following equations will give these additional terms.

Therefore,

$$\sigma_r = \frac{P}{8\pi(1-v)}\left\{ \frac{3r^2 z_1}{R_1^5} - \frac{(1-2v)z_1}{R_1^3} + \frac{(1-2v)z - 12(1-v)h}{R^3}\right.$$
$$\left. - \frac{3r^2 z - 6(7-2v)hz^2 + 24h^2 z}{R^5} - \frac{30hz^2(z-h)}{R^7}\right\} \tag{8.21}$$

$$\sigma_\theta = \frac{P}{8\pi(1-v)}\left\{ -\frac{(1-2v)z_1}{R_1^3}\right.$$
$$\left. + \frac{(1-2v)(z+6h)}{R^3} - \frac{6(1-2v)hz^2 - 6h^2 z}{R^5}\right\} \tag{8.21a}$$

$$\sigma_z = \frac{P}{8\pi(1-v)}\left\{ \frac{3z_1^3}{R_1^5} + \frac{(1-2v)z_1}{R_1^3} - \frac{(1-2v)(z-2h)}{R^3}\right.$$
$$\left. - \frac{3z^3 + 12(2-v)hz^2 - 18h^2 z}{R^5} + \frac{30hz^2(z-h)}{R^7}\right\} \tag{8.21b}$$

$$\tau_{rz} = \frac{Pr}{8\pi(l-v)}\left\{ \frac{3z_1^2}{R_1^5} + \frac{(1-2v)}{R_1^3} - \frac{(1-2v)}{R^3}\right.$$
$$\left. - \frac{3z^2 + 6(3-2v)hz - 6h^2}{R^5} + \frac{30hz^2(z-h)}{R^7}\right\} \tag{8.21c}$$

and

$$\tau_{r\theta} = \tau_{\theta r} = \tau_{\theta z} = \tau_{z\theta} = 0 \tag{8.21d}$$

Mindlin's solution for a horizontal point load also employs the definitions for z_1 and R_1. Now, introduce rectangular co-ordinate system due to the absence of cylindrical symmetry and also replacing r^2 by $x^2 + y^2$, further assuming (without any loss of generality) that the load acts in the x-direction at the point $z = h$, the solution is conveniently stated in terms of Cerrutti's solution, just as the vertical point load was given in terms of Boussinesq's solution.

Therefore, the displacements and stresses to be superposed on Cerrutti's solution are

$$u_x = \frac{P}{16\pi(1-v)G}\left\{ \frac{x^2}{R_1^3} + \frac{(3-4v)}{R_1} - \frac{(3-4v)}{R} \right.$$
$$\left. + \frac{(-x^2+2h(z-h))}{R^3} - \frac{(6hx^2(z-h))}{R^5} \right\} \tag{8.22}$$

$$u_y = \frac{P}{16\pi(1-v)G}\left\{ \frac{xy}{R_1^3} - \frac{xy}{R^3} - \frac{6hxy(z-h)}{R^5} \right\} \tag{8.22a}$$

$$u_z = \frac{P}{16\pi(1-v)G}\left\{ \frac{xz_1}{R_1^3} - \frac{xz+2(3-4v)hz}{R^3} - \frac{6hxz(z-h)}{R^5} \right\} \tag{8.22b}$$

$$\sigma_x = \frac{Px}{8\pi(1-v)}\left\{ \frac{3x^2}{R_1^5} + \frac{(1-2v)}{R_1^3} - \frac{(1-2v)}{R^3} \right.$$
$$\left. - \frac{3x^2-6(3-2v)hz+18h^2}{R^5} - \frac{30hx^2(z-h)}{R^7} \right\} \tag{8.22c}$$

$$\sigma_y = \frac{Px}{8\pi(1-v)}\left\{ \frac{3y^2}{R_1^5} - \frac{(1-2v)}{R_1^3} + \frac{(1-2v)}{R^3} \right.$$
$$\left. - \frac{3y^2-6(1-2v)hz+6h^2}{R^5} - \frac{30hy^2(z-h)}{R^7} \right\} \tag{8.22d}$$

$$\sigma_z = \frac{Px}{8\pi(1-v)}\left\{ \frac{3z_1^2}{R_1^5} - \frac{(1-2v)}{R_1^3} + \frac{(1-2v)}{R^3} \right.$$
$$\left. - \frac{3z^2+6(1-2v)hz+6h^2}{R^5} - \frac{30hz^2(z-h)}{R^7} \right\} \tag{8.22e}$$

$$\tau_{xy} = \frac{Py}{8\pi(1-v)}\left\{ \frac{3x^2}{R_1^5} + \frac{1-2v}{R_1^3} - \frac{1-2v}{R^3} \right.$$
$$\left. - \frac{3x^2-6h(z-h)}{R^5} - \frac{30hx^2(z-h)}{R^7} \right\} \tag{8.22f}$$

$$\tau_{yz} = \frac{Pxy}{8\pi(1-v)}\left\{ \frac{3z_1}{R_1^5} - \frac{3z+6(1-2v)h}{R^5} - \frac{30hz(z-h)}{R^7} \right\} \tag{8.22g}$$

$$\tau_{zx} = \frac{P}{8\pi(1-v)}\left\{\frac{3x^2z_1}{R_1^5} + \frac{(1-2v)z_1}{R_1^3} - \frac{(1-2v)(z-2h)}{R^3}\right.$$
$$\left. - \frac{3x^2z + 6(1-2v)hx^2 - 6hz(z-h)}{R^5} - \frac{30hx^2z(z-h)}{R^7}\right\} \qquad (8.22h)$$

8.7 Applications

The mechanical response of soils and rocks are influenced by a variety of factors such as shape, size and mechanical properties of the individual soil particles, soil structure, the intergranular stresses and stress history, and soil moisture, the degree of saturation and the soil permeability. These factors generally contribute to nonlinear stress–strain phenomena, which is generally irreversible and time dependent. These materials also exhibit anisotropic and non-homogeneous material properties. Thus, any attempt to solve a soil–foundation interaction problem, taking into account all such material properties, is very difficult. In order to obtain meaningful results for practical problems of soil–foundation interaction, it becomes necessary to idealize the behaviour of soil and rocks. The simplest type of idealized soil response assumes linear elastic behaviour.

Two classes of foundation problems can be considered, and they are (1) interactive problems and (2) non-interactive problems. For interactive problems, the elasticity of the foundation plays an important role. For example, a flexible raft foundation supporting a multistorey structure (see Fig. 8.13) interacts with the soil. Using elasticity principles, the deformation of the raft and the deformation of the soil must both obey requirements of equilibrium and must also be geometrically compatible. If a point on the raft is displaced relative to another point, then it can be realized that the bending stresses will develop within the raft and there will be different reactive

Fig. 8.13 A flexible raft foundation supporting a multistorey structure

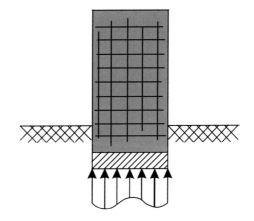

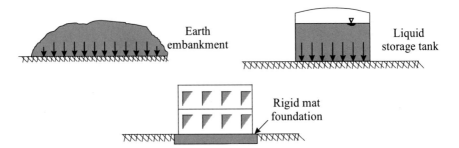

Fig. 8.14 Examples of non-interactive problems

pressures in the soil beneath those points. The response of the raft and the response of the soil are coupled and must be considered together.

Non-interactive problems are those where one can assume the elasticity of the foundation soil as unimportant to the overall response of the soil mass. Some of the non-interactive problems are illustrated in Fig. 8.14. In non-interactive problems, the structural foundation is either very flexible or very rigid when compared with the supporting soil mass. In these problems, it is not necessary to consider the stress–strain response of the foundation. The supporting soil deformations are controlled by the contact exerted by the foundation; however, the response of the soil and the structure is effectively uncoupled.

Non-interactive problems will be much simpler than interactive problems. For problems such as a uniform vertical stress applied at the surface of a homogeneous, isotropic, elastic half-space, one can determine some of the stresses and displacements using Boussinesq's fundamental solutions, which can be obtained by integrating over the region covered by the load. For these problems, there are other methods for finding appropriate solutions, in which specialized mathematical models have been developed that can simulate some of the characteristics of linear elasticity. Winkler model is one such simplest model (Winkler 1867).

Bibliography

1. Barber JR (1992) Elasticity, 2nd edn. Kluwer Publishers, New York
2. Birkhoff' G, MacLane S (1996) A survey of modern algebra, 5th edn. MacMillan, New York
3. Borcsi AP (1965) Elasticity in engineering Mechanics. Prentice Hall, Englewood Cliffs
4. Boussinesq (1878) "Équillibred'élasticité d'un solideisotrope sans pesanteur, supporttantdif-férentspoids". C Rendus Acad Sci Paris 86:1260–1263
5. Cerrutti V (1884) Sulla deformazione di unostratoisotropoindefinito da due pianiparalleli. Atti Acad Nazl Lincei Rend l(4):521–522
6. Davis RO, Selvadurai APS (1996) Elasticity and Geomechanics. Cambridge University Press, Cambridge
7. Fenner RT (1986) Engineering elasticity: Application of numerical and analytical techniques. Harward, Ellis
8. Flamant AA (1892) Sr la repartition des pressionsdans un soliderectangulaire chargé transver-salement. CompteRendu å I 'Acad des Sci 114:1465–1468
9. Fung YC (1969) Introduction to the theory of elasticity. Dover, New York
10. Kazimi SMA (1976) Solid Mechanics. Tata McGraw. Hill, New Delhi
11. Krishna Raju N, Gururaja DR (1997) Advanced mechanics of solids and structures. Narosa Publishing House, New Delhi
12. Landau LD, Lifshitz EM (1989) Theory of Elasticity, 3rd edn. Pergamon, Oxford
13. Martin HS (2009) Elasticity. Academic Press, Burlington, USA
14. Mindlin RD (1936) Force at a point in the interior of a semi-infinite solid. Physics 7:195–202
15. Prescott J (1946) Applied elasticity. Dover, New York
16. Rajagopal KR (1995) Recent advances in elasticity, viscoelasticity and inelasticity. World Scientific, Singapore
17. Sadhu S (1988) Theory' of elasticity. Khanna Publishers, New Delhi
18. Sokolmikoff IS (1956) Mathematical theory of elasticity. McGraw Hill, New York
19. Srinath LS (1982) Advanced Mechanics of Solids. Tata McGraw-Hill, New Delhi
20. Sundara Raja lyengar (1962) Stress distribution in an elastic semi-infinite strip (in German). O.. sterreichischesIngcnieurArchiv 16(3):185–199
21. Timoshenko SP, Goodier JN (1988) Theory' of Elasticity, 2nd edn. McGraw Hill, New York
22. Ugural AC, Fenster SK (1975) Advanced strength and applied elasticity. American Elsevier, New York
23. Valliappan S (1981) Continuum mechanics fundamentals. Oxford and IBH Publishing Company, New Delhi
24. Wang CT (1953) Applied elasticity. McGraw Hill, New York
25. Winkler E (1867) Die lehre von der Elastizitat und Festigkeit. Dominicus, Prague
26. Xu Z (1992) Applied elasticity, Wiley Eastern, New Delhi

T. G. Sitharam and L. Govindaraju, *Theory of Elasticity*,
https://doi.org/10.1007/978-981-33-4650-5

Printed in the United States
by Baker & Taylor Publisher Services